Math Mammoth
Grade 2 Review Workbook

By Maria Miller

Math Mammoth Grade 2 Review Workbook

Contents

Introduction

Math Mammoth Grade 2 Review Workbook is intended to give students a thorough review of second grade math, following the main areas of Common Core Standards for grade 2 mathematics. The book has both topical as well as mixed (spiral) review worksheets, and includes both topical tests and a comprehensive end-of-the-year test. The tests can also be used as review worksheets, instead of tests.

You can use this workbook for various purposes: for summer math practice, to keep the child from forgetting math skills during other break times, to prepare students who are going into third grade, or to give second grade students extra practice during the school year.

The topics reviewed in this workbook are:

- addition and subtraction without regrouping, even and odd numbers, and doubling
- telling time to the five minutes
- addition and subtraction facts within 0-18
- regrouping in addition and subtraction
- geometry topics
- fractions
- three-digit numbers
- measuring
- money
- beginning multiplication

In addition to the topical review worksheets and tests, the workbook also contains many cumulative (spiral) review pages.

The content for these is taken from *Math Mammoth Grade 2 Complete Curriculum*, so naturally this workbook works especially well to prepare students for grade 3 in Math Mammoth. However, the content follows a typical study for grade 2, so this workbook can be used no matter which math curriculum you follow.

Please note this book does not contain lessons or instruction for the topics. It is not intended for initial teaching. It also will not work if the student needs to completely re-study these topics (the student has not learned the topics at all). For that purpose, please consider *Math Mammoth Grade 2 Complete Curriculum*, which has all the necessary instruction and lessons.

I wish you success with teaching math!

Maria Miller, the author

Some Old, Some New

1. Add. The problems in each box are similar.

a.	b.	c.	d.
$51 + 7 = 58$	$46 + 3 = 49$	$72 + 5 = 77$	$35 + 5 = 40$
$81 + 7 = 88$	$96 + 3 = 99$	$32 + 5 = 37$	$95 + 5 = 100$

2. Subtract. The problems in each box are similar.

a.	b.	c.	d.
$49 - 5 = 44$	$29 - 3 = 26$	$60 - 7 = 53$	$38 - 4 = 34$
$89 - 5 = 84$	$69 - 3 = 66$	$80 - 7 = 73$	$78 - 4 = 74$

3. **a.** How much would three $20 shirts cost together? $60

   b. Mike went to a yard sale and bought a desk for $32,
   a toy car for $1, a plant for $2, and some thread for $4.
   What was his total bill?

   $$\begin{array}{r} 32 \\ 1 \\ 2 \\ 4 \\ \hline 39 \end{array}$$

4. Add and subtract whole tens.

a.	b.	c.	d.
$21 + 40 = 61$	$40 + 23 = 63$	$72 - 50 = 22$	$89 - 30 = 59$
$56 + 30 = 86$	$20 + 78 = 98$	$66 - 40 = 26$	$45 - 20 = 25$

5. Use letters from the word **W O N D E R F U L** to make two new words.

___	___	___	___		___	___	___	___
1st	5th	9th	9th		4th	2nd	3rd	5th

6. Fill in the missing numbers. The four problems form a fact family.

a. 2 + [8] = 10

   [8] + 2 = 10

   10 − 2 = [8]

   10 − [8] = 2

b. 7 + 2 = 9

   2 + 7 = 9

   9 − 7 = 2

   9 − 2 = 7

c. 3 + 5 = 8

   5 + 3 = 8

   8 − 3 = 5

   8 − 5 = 3

7. The total is missing from the subtraction sentence. Solve.

a. [16] − 8 = 8

b. [9] − 5 = 4

c. [60] − 30 = 30

8. Circle the even numbers.

72   31   59   60   8

9. Divide the dots into two EQUAL groups. Find half of the total.

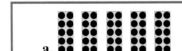

a. $\frac{1}{2}$ of 50 is _25_.

b. $\frac{1}{2}$ of 88 is _44_.

c. $\frac{1}{2}$ of 46 is _23_.

10. Two boys divided evenly 18 toy cars.
    How many did each boy get?   9 cars

11. Mrs. Smith used half of her potatoes to make mashed potatoes.
    Now she has 13 potatoes left. How many did she have at first?   26

12. Mary has 13 colored pencils and Tina has twice as many.
    How many colored pencils do the girls have together?   39

   13 + 13
   ‾‾‾‾‾‾‾
     26
   ‾‾‾‾‾‾‾
     39

7

Some Old, Some New - Test

1. Add and subtract.

a.	b.	c.	d.
$58 + 2 =$ _60_	$33 + 4 =$ _37_	$50 + 6 =$ _56_	$45 + 40 =$ _85_
$58 - 6 =$ _52_	$94 - 4 =$ _90_	$65 - 30 =$ _35_	$98 - 70 =$ _28_

2. Color.

 **a.** The fourth flower from the left.	 **b.** The sixth flower from the right.

3. Fill in the missing numbers. The four problems form a fact family.

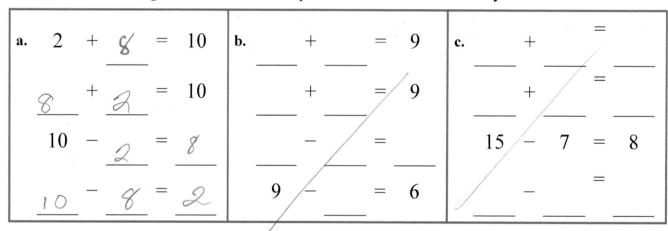

a.
$2 + 8 = 10$
$8 + 2 = 10$
$10 - 2 = 8$
$10 - 8 = 2$

b.
$\_ + \_ = 9$
$\_ + \_ = 9$
$\_ - \_ = \_$
$9 - \_ = 6$

c.
$\_ + \_ = \_$
$\_ + \_ = \_$
$15 - 7 = 8$
$\_ - \_ = \_$

4. You read 23 pages in a story book. Your friend Sally read double that many pages. How many pages did Sally read?

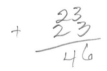

	2	3
+	2	3
	4	6

5. Are these numbers even or odd? Mark an "X".

Number	Even?	Odd?
4	even	
10	even	

Number	Even?	Odd?
9		odd
16	even	

Number	Even?	Odd?
11		odd
18	even	

Clock Review

1. Write the time with *hours:minutes*, and using "past", "till", "half past" or "o'clock".

a. 2 : 50 _____ till _____	**b.** 4 : 25 _____ past _____	**c.** 9 : 55 _____	**d.** 11 : 05 _____
e. 3 : 40 _____ till _____	**f.** 7 : 25 _____	**g.** 5 : 30 _____	**h.** 12 : o'clock _____

2. Write the later time.

Time now	2:30	6:55
5 min. later	2 :35	7 :00

Time now	9:05	5:40
10 min. later	9 :15	5 :50

3. Father starts his work at 9 AM, and leaves to go back home at 5 PM.
   How many hours is his work day?   8 hours

4. Little Emily goes to kindergarten at 8 AM and stays there four hours.
   At what time does her kindergarten class end?   12 o'clock

5. Jonah goes to the chess club every Thursday. He went today, March 17.
   What is the date when he goes next time?   March 24th

6. Emily got the August issue of a magazine in the mail. The next
   magazine comes in three months. What month will that be?   November

9

Clock Test

1. Write the time with hours:minutes, and using "past", "till", "half past" or "o'clock".

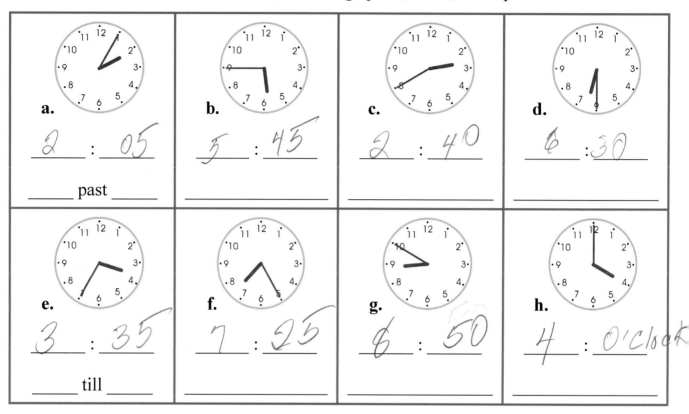

a. _2_ : _05_

_____ past _____

b. _5_ : _45_

_____

c. _2_ : _40_

_____

d. _6_ : _30_

_____

e. _3_ : _35_

_____ till _____

f. _7_ : _25_

_____

g. _6_ : _50_

_____

h. _4_ : _O'clock_

_____

2. Write the later time.

Time now	3:50	7:25
5 minutes later	3:55	7:30

Time now	9 AM	12 noon
1 hour later	10 am	1 pm

3. Write the time using the **hours:minutes** way.

a. 20 past 4	**b.** 15 past 11	**c.** 15 till 12	**d.** 25 till 7
4 : 20	11 : 15	12 : 15	7 : 25

4. How many hours pass?

from	5 AM	8 AM	2 AM	10 AM	11 AM
to	12 noon	2 PM	3 PM	10 PM	6 PM
hours	7 hours	6 hours	11 hours	12 hours	7 hours

10

Mixed Review 1

1. Solve the problems. Fill in the doubles chart. *It has a pattern!* Find it!

   a. It will take Alex 16 hours to clean the park. He did half of that yesterday. How many hours will he still have to work?

   8 hours

   b. What is double 12?

   24

   $30/2 = 15$ each

   c. Ava and Emma divided evenly $30. Then Emma bought a gift for $6. How much money does Emma have now?

   $15 - 6 = 9$

   d. Eddie has saved $20. That is just half of what he needs to buy a train set. How much does the train set cost?

   $ 40

   $10 + 10 =$ _20_

   $15 + 15 =$ _30_

   $20 + 20 =$ _40_

   $25 + 25 =$ _50_

   $30 + 30 =$ _60_

   $35 + 35 =$ _70_

   $40 + 40 =$ _80_

2. Add and subtract whole tens.

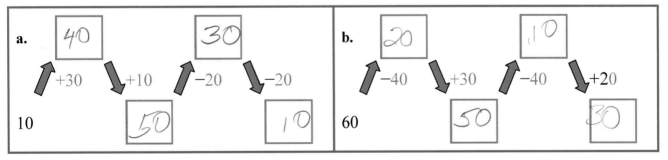

3. Write the time using *hours:minutes*.

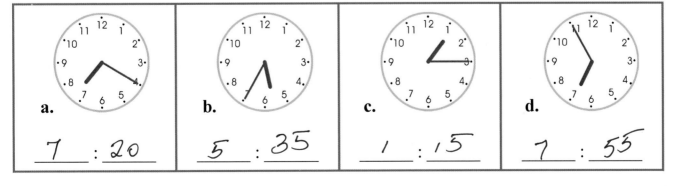

   a. 7 : 20

   b. 5 : 35

   c. 1 : 15

   d. 7 : 55

4. **a.** Anne began watching a film about sea animals at 20 till 4. She stopped watching it at 15 past 4. Write those two times in the *hours:minutes* way.

___3___ : __40__ and ___4___ : __15__

   b. Jim began walking his dog at 11 AM and stopped at noon.
   How long did he walk his dog?

   one hour

   c. Bill's rooster crowed for half an hour, starting at 5 AM.
   At what time did it stop?

   5:30 am

5. Fill in the missing numbers. The four problems form a fact family.

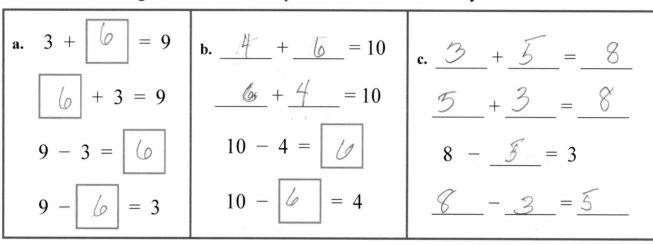

a. $3 + \boxed{6} = 9$

$\boxed{6} + 3 = 9$

$9 - 3 = \boxed{6}$

$9 - \boxed{6} = 3$

b. $\underline{4} + \underline{6} = 10$

$\underline{6} + \underline{4} = 10$

$10 - 4 = \boxed{6}$

$10 - \boxed{6} = 4$

c. $\underline{3} + \underline{5} = \underline{8}$

$\underline{5} + \underline{3} = \underline{8}$

$8 - \underline{5} = 3$

$\underline{8} - \underline{3} = \underline{5}$

6. Find the letters, and find out what Bob got for his birthday.

The second row from the top,
the second letter from the left.   ____

The fourth row from the top,
the fifth letter from the left.   ____

The first row from the top,
the fifth letter from the right.   ____

The fifth row from the bottom,
the second letter from the right.   ____

The 1st row from the bottom,
the 1st letter from the left.   ____

The sixth row from the top,
the third letter from the right.   ____

The 3rd row from the top,
the 2nd letter from the left.   ____

E	S	H	A	B	G	P
B	A	E	N	I	V	S
W	E	K	P	T	O	F
J	D	A	U	-	W	M
Y	K	Z	N	Y	I	C
U	D	T	S	S	Q	R
R	T	H	A	V	E	L

Addition and Subtraction Facts Within 0-18 Review

1. Here are the 20 addition facts with single-digit numbers where the sum is between 10 and 20. Connect the problems to the right answer.

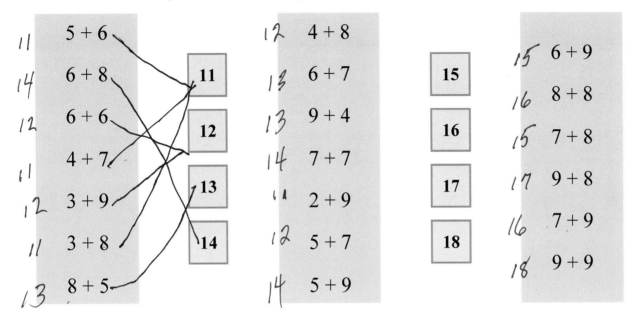

2. Draw a line to connect the problems that are in the same fact family. You do not need to write the answers.

13 − 7 = ▢		12 − 5 = ▢		15 − 7 = ▢
7 + ▢ = 15		11 − 8 = ▢		13 − 6 = ▢
11 − 3 = ▢		9 + ▢ = 17		5 + ▢ = 14
8 + ▢ = 17		15 − 8 = ▢		17 − 8 = ▢
14 − 5 = ▢		6 + ▢ = 13		3 + ▢ = 11
7 + ▢ = 12		9 + ▢ = 14		▢ + 5 = 12

3. Find the differences.

a. Between 80 and 87 _____	**b.** Between 45 and 2 _____
c. Between 15 and 8 _____	**d.** Between 13 and 4 _____

4. Find the missing numbers.

a. $8 + \boxed{} = 15$	**b.** $7 + \boxed{} = 14$	**c.** $6 + \boxed{} = 13$
d. $13 - \boxed{} = 5$	**e.** $14 - \boxed{} = 8$	**f.** $15 - \boxed{} = 9$
g. $11 - 6 = \boxed{}$	**h.** $12 - 7 = \boxed{}$	**i.** $12 - 4 = \boxed{}$

5. Find the missing steps.

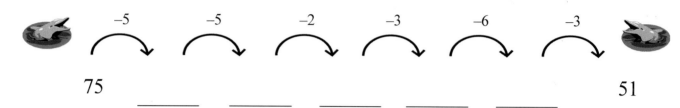

75 _____ _____ _____ _____ _____ 51

6. **a.** You have an *odd* number of cookies and so does your friend. You put your cookies together and share them. Can you share them evenly or not?

Cookies you have	Cookies your friend has	Together we have	even/odd	Can you share evenly?
3	5			
5	9			
9	3			
9	7			

b. You have an *odd* number of cookies and your friend has an *even* number of cookies. You put your cookies together and share them. Can you share them evenly or not?

Cookies you have	Cookies your friend has	Together we have	even/odd	Can you share evenly?
5	6			
7	8			
9	4			
1	12			

7. Solve the puzzle. What happened to the teddy bear in the desert?

| $5+9$ | $7+8$ | | $13-8$ | $2+9$ | $10+5$ | | $9+7$ | $4+7$ | $9+6$ |

___ ___   ___ ___ ___   ___ ___ ___

| $7+7$ | $13-6$ | | $19-4$ | $11+5$ | $13-7$ | | $3+13$ | $11-5$ | $13-4$ | $6+9$ |

___ ___   ___ ___ ___   ___ ___ ___ ___

Key:

A	E	I	O	G	H	T	W	N
9	6	14	11	5	16	15	8	7

8. Solve the word problems.

a. Jack has 13 tennis balls and Jane has 20.
How many more does Jane have than Jack?

b. Emma has three more flowers than Sofia. If Emma
has 14 flowers, how many does Sofia have?

c. In a chess game, Jacob has 2 more pawns than Anna.
If Anna has five pawns, how many does Jacob have?

d. You have $20, and you want to buy a Lego set that costs $28.
How many dollars do you still need to save?

Later, a neighbor pays you $2 for helping rake leaves.
How much more money do you need after that?

e. In a board game, you need to move 18 more squares to get to the end
of the game. You roll 6 and 5 on two dice and move that many squares.
Now how many more squares are there to the end?

What kind of numbers on the two dice would get you to the end?

Addition and Subtraction Facts Within 0-18 Test

1. Add and find the missing numbers.

a. $9 + 6 =$ _____ $9 + 4 =$ _____	**b.** $8 + 9 =$ _____ $7 + 5 =$ _____	**c.** $7 +$ _____ $= 14$ $7 +$ _____ $= 16$
d. $9 +$ _____ $= 12$ $9 +$ _____ $= 18$	**e.** $8 + 5 =$ _____ $6 + 7 =$ _____	**f.** $6 + 8 =$ _____ $8 + 7 =$ _____

2. Subtract. For each problem write a corresponding addition fact.

a. $14 - 5 =$ _____ _____ $+$ _____ $=$ _____	**b.** $11 - 8 =$ _____ _____ $+$ _____ $=$ _____	**c.** $17 - 8 =$ _____ _____ $+$ _____ $=$ _____

3. Write $<$, $>$, or $=$.

   a. $7 + 9$ ☐ $8 + 8$ **b.** $40 - 5$ ☐ $40 - 8$ **c.** $\frac{1}{2}$ of 20 ☐ $\frac{1}{2}$ of 18

4. Subtract.

a. $11 - 6 =$ _____ $17 - 9 =$ _____	**b.** $16 - 8 =$ _____ $14 - 8 =$ _____	**c.** $13 - 6 =$ _____ $15 - 8 =$ _____

5. Solve.

a. Annie has 7 more teddy bears than Jason. Jason has 9.   How many does Annie have?

b. You have saved \$7 to buy a book that costs \$14.
    Then, Grandma gives you \$5.
    How much money do you still need?

Mixed Review 2

1. Write the time.

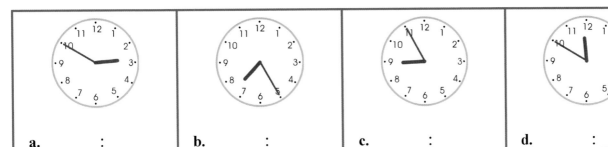

a. _____ : _____   b. _____ : _____   c. _____ : _____   d. _____ : _____

2. Write the time that the clock shows, and the time 5 minutes later.

	a.	b.	c.	d.
	_____ : _____	_____ : _____	_____ : _____	_____ : _____
5 min. later →	_____ : _____	_____ : _____	_____ : _____	_____ : _____

3. How many minutes pass? Subtract (or figure out the difference).

from	2:25	2:20	7:00	11:30	6:05
to	2:35	2:40	7:15	11:50	6:15
minutes	10 minutes				

4. Ashley shared 18 raisins and 12 almonds equally with her brother.

How many raisins did each child get?

How many almonds?

5. Write each number as a double of some other number.

a. 10 = _____ + _____	b. 16 = _____ + _____	c. 40 = _____ + _____

6. Fill in the missing numbers for this subtraction "journey".

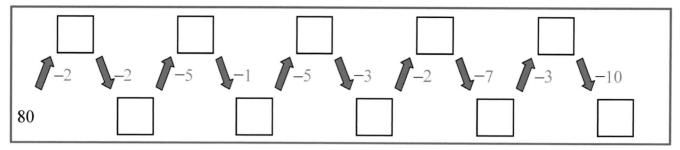

7. Solve the problems.

a. Mary ate 20 strawberries and Isabella ate half that many.

How many did Isabella eat?

How many did the girls eat together?

b. Kyle used half of his money to buy a toy car. Now he has $10 left.
How much money did Kyle have at first?

c. Jane ate 10 strawberries more than what Jonathan ate.
Jonathan ate 12 strawberries. How many did Jane eat?

d. Emily is 30 years old, and Hannah is 4 years old.
How many years older is Emily than Hannah?

e. Ann had 12 toy cars, and Judith had 10. Then Ann got two more cars.

Now who has more cars?

How many more?

f. Jacob has $6 and Jim has $7 more than Jacob.
How much money does Jim have?

Mixed Review 3

1. Add and find the missing addends.

a.	b.	c.	d.
$6 + 7 = $ _____	$9 + 7 = $ _____	$5 + $ _____ $= 14$	$8 + $ _____ $= 15$
$8 + 9 = $ _____	$5 + 8 = $ _____	$8 + $ _____ $= 16$	$7 + $ _____ $= 14$

2. How many hours is it?

from	9 AM	6 AM	11 AM	12 AM	10 AM
to	1 PM	8 PM	4 PM	12 PM	2 PM
hours					

3. **a.** How many Tuesdays are there in January?
   (See the calendar on the right.)

January

Su	Mo	Tu	We	Th	Fr	Sa
		1	2	3	4	5
6	7	8	9	10	11	12
13	14	15	16	17	18	19
20	21	22	23	24	25	26
27	28	29	30	31		

   b. Jane visits her parents every third Sunday of the
   month. What day will she visit them in January?

4. Solve.

a. Joyce practices playing the piano for 2 hours.
   She stopped practicing at 2 PM. When did she *start* practicing?

b. Grandma sleeps 1/4 of the day's hours. (One day has 24 hours.)
   How many hours does Grandma sleep each day?

5. Add and subtract whole tens.

a. $77 + 20 = $ _____	b. $18 + 50 = $ _____	c. $54 + 40 = $ _____
$64 - 30 = $ _____	$43 - 20 = $ _____	$98 - 60 = $ _____

6. Add more. Find the difference.

a. $18 + \underline{\hspace{1cm}} = 22$	**b.** $75 + \underline{\hspace{1cm}} = 80$	**c.** $56 + \underline{\hspace{1cm}} = 59$
d. The difference of 8 and 12 is $\underline{\hspace{1cm}}$.	**e.** The difference of 43 and 49 is $\underline{\hspace{1cm}}$.	**f.** The difference of 21 and 30 is $\underline{\hspace{1cm}}$.

7. Subtract. Think about the difference.

a. $85 - 80 = \underline{\hspace{1cm}}$	**b.** $76 - 71 = \underline{\hspace{1cm}}$	**c.** $20 - 17 = \underline{\hspace{1cm}}$
$46 - 42 = \underline{\hspace{1cm}}$	$99 - 89 = \underline{\hspace{1cm}}$	$70 - 67 = \underline{\hspace{1cm}}$

8. For each addition, write a matching subtraction (using the same numbers).

a. $8 + \boxed{} = 14$ $\underline{\hspace{1cm}} - \underline{\hspace{1cm}} = \boxed{}$	**b.** $5 + \boxed{} = 14$ $\underline{\hspace{1cm}} - \underline{\hspace{1cm}} = \boxed{}$	**c.** $6 + \boxed{} = 12$ $\underline{\hspace{1cm}} - \underline{\hspace{1cm}} = \boxed{}$

9. Subtract.

a. $12 - 7 = \underline{\hspace{1cm}}$	**b.** $14 - 8 = \underline{\hspace{1cm}}$	**c.** $11 - 6 = \underline{\hspace{1cm}}$	**d.** $15 - 7 = \underline{\hspace{1cm}}$
$17 - 9 = \underline{\hspace{1cm}}$	$12 - 8 = \underline{\hspace{1cm}}$	$13 - 8 = \underline{\hspace{1cm}}$	$14 - 9 = \underline{\hspace{1cm}}$
$11 - 8 = \underline{\hspace{1cm}}$	$13 - 7 = \underline{\hspace{1cm}}$	$16 - 9 = \underline{\hspace{1cm}}$	$15 - 9 = \underline{\hspace{1cm}}$

10. Detective Cole was a math sleuth. He was out to get the fact family. He had found number 13, but two numbers were missing. Help him find the fact family!

He found a clue under the couch: "Look in the cookie jar!" In the cookie jar there were half a dozen cookies left. Cole said, "That's one of my missing numbers!"

Can you figure out the other missing number now? Then, write the fact family.

$\underline{\hspace{0.7cm}} + \underline{\hspace{0.7cm}} = \underline{\hspace{1cm}}$ $\underline{\hspace{1cm}} - \underline{\hspace{0.7cm}} = \underline{\hspace{0.7cm}}$

$\underline{\hspace{0.7cm}} + \underline{\hspace{0.7cm}} = \underline{\hspace{1cm}}$ $\underline{\hspace{1cm}} - \underline{\hspace{0.7cm}} = \underline{\hspace{0.7cm}}$

The case is solved!

Regrouping in Addition Review

1. Add *part-by-part*. Break one of the numbers into its tens and ones in your mind.

a. $17 + 10 =$ _____	**b.** $16 + 20 =$ _____	**c.** $50 + 14 =$ _____
$42 + 10 =$ _____	$67 + 20 =$ _____	$30 + 45 =$ _____

2. Add.

a. $27 + 8 =$ _____	**b.** $18 + 9 =$ _____	**c.** $5 + 87 =$ _____
$54 + 7 =$ _____	$73 + 8 =$ _____	$7 + 88 =$ _____

3. Add by adding tens and ones separately.

a. $\qquad 36 + 22$ $30 + 20 + 6 + 2$ _____ + _____ = _____	**b.** $\qquad 72 + 18$ $70 + 10 + 2 + 8$ _____ + _____ = _____
c. $\qquad 54 + 37$ $50 + 30 \ + 4 + 7$ _____ + _____ = _____	**d.** $\qquad 24 + 55$ _____ + _____ + _____ _____ + _____ = _____

4. Solve the problems.

a. Diane and Ted picked fruit for Mr. Mohan. Diane earned $25 and Ted earned double that. How much did Ted earn?   How much did the two earn together?
b. Emily has 24 flower plants in her yard. Leah has half that many. How many flower plants does Leah have?

5. Add.

a. 43
  + 28
  ____

b. 33
  + 39
  ____

c. 24
  + 47
  ____

d. 23
  + 38
  ____

e. 55
  + 17
  ____

f. 38
    13
  + 42
  ____

g. 39
    10
  + 46
  ____

h. 41
    44
  + 36
  ____

i. 38
     7
    49
  + 23
  ____

j. 27
    36
    19
  + 35
  ____

6. Solve.

a. Naomi bought some potatoes for $18, onions for $15, and meat for $40. What was the total cost?

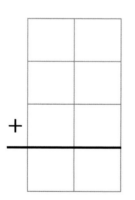

b. If you buy three chairs for $34 each, what is the total bill?

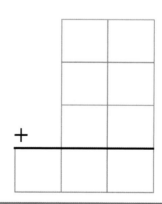

c. Anna has 29 stickers and so does Betty. Ruth has 22 and Judy has 26. How many stickers are there total?

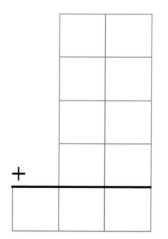

d. Andy had $47 in his wallet. He earned $15 by selling lemonade. Now can he buy a remote-controlled toy car for $65?

If yes, how much money would have left after buying it?

If not, how much more money would he need?

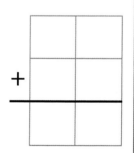

Regrouping in Addition Test

1. Add.

a.
```
  3 9
+ 4 6
─────
```

b.
```
  8 3
  1 4
+ 2 5
─────
```

c.
```
  4 6
    8
+ 3 3
─────
```

d.
```
  3 8
+ 2 3
─────
```

e.
```
  2 4
    7
  5 8
+ 1 5
─────
```

2. Add.

a. 52 + 7 = _____

18 + 5 = _____

b. 67 + 6 = _____

27 + 8 = _____

c. 88 + 5 = _____

43 + 8 = _____

3. Add mentally.

a. 2 + 6 + 8 + 7 = _____

5 + 7 + 4 + 8 = _____

b. 42 + 2 + 10 + 5 = _____

30 + 30 + 9 + 7 = _____

4. Solve the problems.

a.
Mary worked for 28 hours and Jill worked for 13. How many more hours did Mary work than Jill?

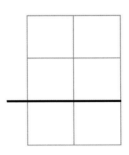

b.
Find the total cost if you buy a stuffed animal for $12 and two books for $17 each.

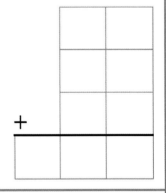

c. Fifteen birds were perching on a tree. Then, nine more birds flew in, and two birds flew away. How many birds are in the tree now?

Mixed Review 4

1. Find one-half and double of the given numbers.

a. $\frac{1}{2}$ of 6 is _____.	b. $\frac{1}{2}$ of 10 is _____.	c. $\frac{1}{2}$ of 8 is _____.
d. Double 6 is _____	e. Double 10 is _____	f. Double 8 is _____

2. Find the number that goes into the shape.

a. $73 + \bigcirc = 80$	b. $78 + \bigcirc = 98$	c. $96 - \bigcirc = 56$
d. $\bigcirc + 92 = 100$	e. $\bigcirc - 20 = 5$	f. $\bigcirc - 50 = 41$

3. Draw a line to connect the problems that are in the same fact family.
   (You don't need to solve them.)

$13 - 8 = \square$	$13 - 6 = \square$	$15 - 6 = \square$
$6 + \square = 15$	$11 - 2 = \square$	$13 - 5 = \square$
$11 - 9 = \square$	$7 + \square = 15$	$5 + \square = 11$
$8 + \square = 15$	$15 - 9 = \square$	$15 - 8 = \square$
$11 - 5 = \square$	$5 + \square = 13$	$9 + \square = 11$
$7 + \square = 13$	$6 + \square = 11$	$\square + 6 = 13$

4. Subtract.

a.	b.	c.	d.
$15 - 9 = $ _____	$13 - 9 = $ _____	$14 - 8 = $ _____	$15 - 7 = $ _____
$13 - 6 = $ _____	$14 - 7 = $ _____	$16 - 8 = $ _____	$13 - 5 = $ _____

5. Write the time.

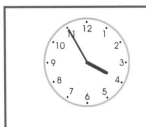			
a. _____ : _____	b. _____ : _____	c. _____ : _____	d. _____ : _____

6. Write the time 10 minutes later than what the clocks show in the previous exercise.

a. _____ : _____	b. _____ : _____	c. _____ : _____	d. _____ : _____

7. Solve the problems.

a. In a game, Kathy got 14 points, Shaun got double that many points, and Abigail got 10 more points than Kathy.
Who got the most points?

How many points was that?

b. You are 8 years old and your brother is double your age.
How many years older is your brother than you?

c. Emma got 7 points more in a game than Matthew.
Matthew got 31 points. How many points did Emma get?

d. One volleyball costs $26 and another costs $6 more than that.
How much does the other volleyball cost?

e. There were 7 more birds in the oak tree than in the birch tree.
If the oak tree had 15 birds, how many birds were in the birch tree?

f. Edith has 12 markers and Judith has 6. They put together their markers and share them evenly. How many does each girl get?

Mixed Review 5

1. Solve the problems.

> **a.** One-half of the boys in the class are studying math.
> The other seven boys are reading.
> How many boys are in the class?

> **b.** Mark has $8. Andy has double that much money.
> How much money does Andy have?
> How much money do the two boys have together?

2. Add mentally.

a. $28 + 4 =$ _____	**b.** $39 + 9 =$ _____	**c.** $44 + 5 + 4 =$ _____
$28 + 40 =$ _____	$30 + 29 =$ _____	$7 + 8 + 9 + 4 =$ _____

3. A few years ago a small camera cost $67.
   Now it has doubled in price.
   How much does it cost now?

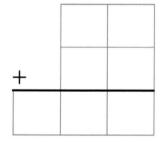

4. Write the time using "past", "till", "half past", or "o'clock".

a. 7:25 _____	**b.** 5:10 _____
c. 5:50 _____	**d.** 12:40 _____
e. 12:30 _____	**f.** 11:00 _____

5. Write the numbers so that ones and tens are in their own columns. Add.

   a. 44 + 37   **b.** 9 + 26   **c.** 26 + 8 + 47   **d.** 25 + 57 + 38

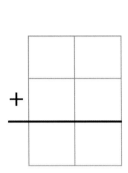 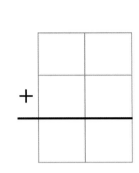 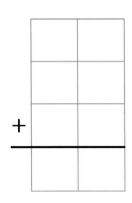 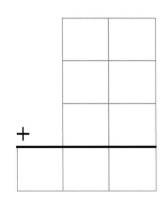

6. Chris made six cards for his party. He put the cards at the plate of each guest.
   It took Chris 10 minutes to make one card.

   a. How long did it take Chris to make all 6 cards?

   b. If he started making the cards at noon, at what time did he finish his project?

7. Fill in the missing numbers.

a. $24 + 8 = \bigcirc$	**b.** $16 - 7 = \bigcirc$	**c.** $17 - 9 = \bigcirc$
d. $\bigcirc - 6 = 5$	**e.** $\bigcirc - 20 = 7$	**f.** $\bigcirc - 5 = 31$

8. You bought three stools for $18
   each and some towels for $25.
   Find the total cost.

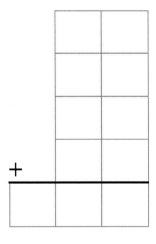

Geometry and Fractions Review

1. Connect the dots. Use a ruler!
   What shape do you get?

   _____

2. Choose one corner of your shape.
   Now draw a line (with a ruler)
   from that corner to some other
   corner so that you will divide the
   shape into a <u>triangle</u> and a <u>pentagon</u>.

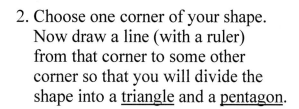

3. Draw a square in the grid that
   has 4 little squares inside.

4. Draw a rectangle in the grid that
   has 18 little squares inside.

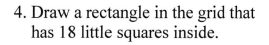

5. What is this shape called? _____

   How many *faces* does it have? _____

   What shape are the faces? _____

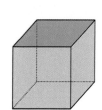

6. Sarah put together these two triangles. What new shape did she get?

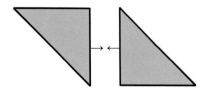

7. Label the pictures as *box*, *cylinder*, *pyramid*, or *cone*.

    a.

    b.

c.

_____   _____   _____

8. Color the whole shape. Then write 1 whole as a fraction. Lastly, read what you wrote with numbers.

   a. $1 = \dfrac{\quad}{\quad}$ b. $1 = \dfrac{\quad}{\quad}$

9. Divide the shapes into two, three, or four equal parts so that you can color the fraction.

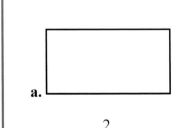
a.
$\dfrac{2}{4}$

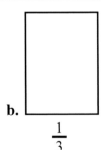
b.
$\dfrac{1}{3}$

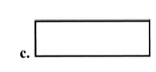
c.
$\dfrac{2}{3}$

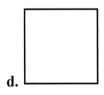
d.
$\dfrac{2}{2}$

10. Color. Then compare and write $<$, $>$ or $=$. Which is more "pie" to eat?

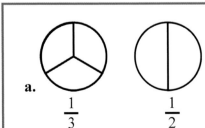
a. $\dfrac{1}{3}$ $\dfrac{1}{2}$

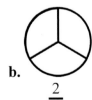
b. $\dfrac{2}{3}$ $\dfrac{3}{4}$

c.   1 whole $\dfrac{3}{4}$

Geometry and Fractions Test

1. Join the dots with lines. Use a ruler. Write the name of the shape you get.

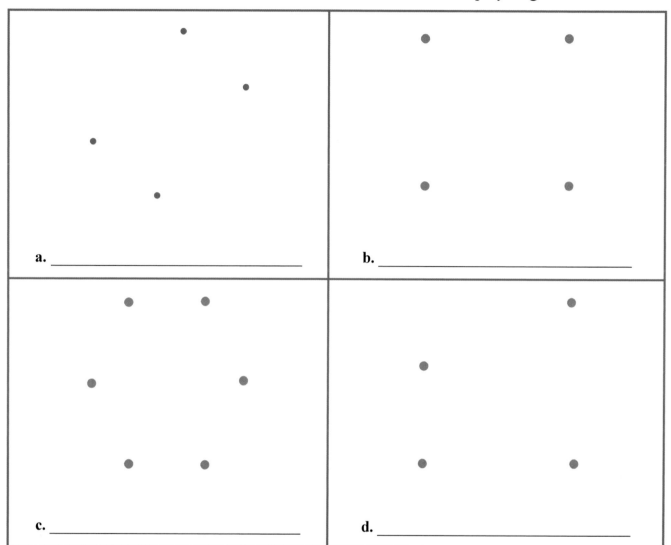

a. _____

b. _____

c. _____

d. _____

2. Draw a rectangle so it has a certain number of little squares inside.

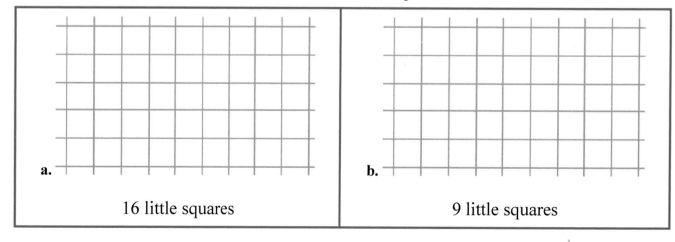

a. 16 little squares

b. 9 little squares

3. Design a pattern with rectangles and/or squares.

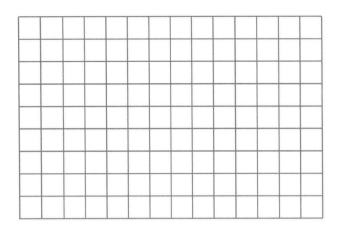

4. Write the fraction.

a. ——   b. ——   c. ——   d. ——

5. Divide these shapes. Then color as you are asked to.

a.	b.	c.	d.
Divide this into thirds. Color $\frac{1}{3}$.	Divide this into halves. Color $\frac{2}{2}$.	Divide this into halves. Color $\frac{1}{2}$.	Divide this into fourths. Color $\frac{3}{4}$.

6. Color. Then compare and write < , > , or = .

a. 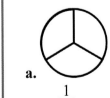	b. 	c. 
$\frac{1}{3}$ $\frac{2}{5}$	$\frac{5}{6}$ $\frac{3}{4}$	$\frac{3}{3}$ $\frac{4}{4}$

Mixed Review 6

1. Find the differences.

a. The difference of 100 and 95 _____	**b.** The difference of 40 and 20 _____
c. The difference of 16 and 8 _____	**d.** The difference of 56 and 4 _____

2. Subtract. Think of the difference.

a. $25 - 22 =$ _____	**b.** $76 - 71 =$ _____	**c.** $51 - 49 =$ _____

3. Find the missing numbers.

a. $14 - \boxed{} = 5$	**b.** $13 - \boxed{} = 8$	**c.** $16 - \boxed{} = 9$
d. $\boxed{} - 6 = 6$	**e.** $\boxed{} - 7 = 4$	**f.** $\boxed{} - 4 = 9$

4. Add. Compare the problems.

a. $8 + 3 =$ _____ $18 + 3 =$ _____	**b.** $6 + 6 =$ _____ $86 + 6 =$ _____	**c.** $8 + 7 =$ _____ $48 + 7 =$ _____
d. $46 + 7 =$ _____	**e.** $47 + 9 =$ _____	**f.** $88 + 5 =$ _____

5. Add. Regroup the ones to make a new ten.

a.
```
   6 4
   1 5
 + 2 5
 ------
```

b.
```
   4 7
   2 7
 + 2 3
 ------
```

c.
```
   1 3
   5 6
 + 2 6
 ------
```

d.
```
   1 5
   2 6
   4 7
 + 1 9
 ------
```

e.
```
   2 7
     9
   3 5
 + 2 5
 ------
```

6. Find how much the things cost together.

a.	b.	c.
a fishing rod, $38 baits, $9 bucket, $8	skis, $79 jacket, $22 socks, $11	four chairs, $29 each

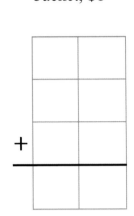 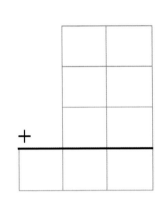 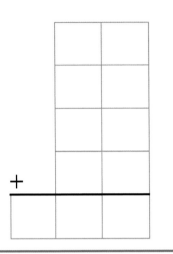

7. Add four numbers. Look at the example.

a. $8 + 8 + 2 + 8$ $= 16 + 10$ $= 26$	b. $9 + 5 + 5 + 8$ $= \underline{\quad} + \underline{\quad}$ $= \underline{\quad}$	c. $6 + 7 + 3 + 5$ $= \underline{\quad} + \underline{\quad}$ $= \underline{\quad}$
d. $7 + 7 + 8 + 8$ $= \underline{\quad}$	e. $9 + 4 + 4 + 7$ $= \underline{\quad}$	f. $6 + 4 + 4 + 9$ $= \underline{\quad}$

8. Solve the problems. You need to add or subtract.

a. One bike costs $78, and another costs $23 more than the first. Find the price of the second bike.	b. One shirt costs $29 and another costs $15. How much more does the first shirt cost than another?	c. You bought both shirts in problem (b). How much did they cost together?

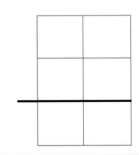 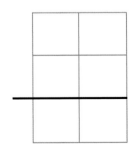 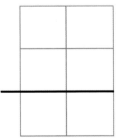

Mixed Review 7

1. Write two additions and two subtractions for each picture. The box with a "T" is a ten.

a. T / T   and   T T / T T

_____ + _____ = _____

_____ + _____ = _____

_____ − _____ = _____

_____ − _____ = _____

b. T T ●●●●   and   (dots)

_____ + _____ = _____

_____ + _____ = _____

_____ − _____ = _____

_____ − _____ = _____

2. Add.

   a. 29 + 90   b. 93 + 46   c. 24 + 35 + 48   d. 22 + 47 + 9

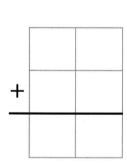 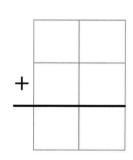 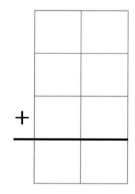 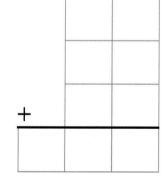

3. Solve.

a. Choose the letters from the given word to make a new word.

   M A M M A L

   ____ ____ ____ ____ ____
   6th   6th   5th   3rd   2nd

b. Put the letters in the given order to make a new word.

   L   O   I   D   R   N   A
   7th   1st   4th   3rd   2nd   5th   6th

   O ____ ____ ____ ____ ____ ____

4. Write the time that the clock shows, and the time 5 minutes later.

	a. _____ : _____	b. _____ : _____	c. _____ : _____	d. _____ : _____
5 min. later →	_____ : _____	_____ : _____	_____ : _____	_____ : _____

5. A flower vase has 15 flowers. Some are red, some are blue, and some are yellow. Four of the flowers are red and five are yellow. How many are blue?

6. Write $<$, $>$, or $=$. You can often compare without calculating!

a. $8 + 8$ ☐ $9 + 8$ b. $30 - 8$ ☐ $30 - 9$ c. $\frac{1}{2}$ of 16 ☐ 16

d. $35 + 7$ ☐ $35 + 8$ e. $40 - 6$ ☐ $40 - 9$ f. $14 - 7$ ☐ $16 - 8$

7. Add by adding tens and ones separately.

a. $36 + 22$ $30 + 20 + 6 + 2$ _____ + _____ = _____	b. $72 + 18$ $70 + 10 + 2 + 8$ _____ + _____ = _____
c. $54 + 37$ _____ + _____ = _____	d. $24 + 55$ _____ + _____ = _____

8. Count by 10s and 50s, and fill in the grids.

a.	464	474							

b.	400	450							

Three-Digit Numbers Review

1. **a.** Write the number shown by the image: _____

    b. Write the number that is 1 more than the number in the image: _____

    c. Write the number that is 10 more than the number in the image: _____

    d. Write the number that is 100 more than the number in the image: _____

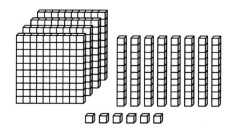

2. Write the numbers that come before and after the given number.

    a. _____ , 179 , _____

    b. _____ , 201 , _____

    c. _____ , 800 , _____

    d. _____ , 917 , _____

3. Write with numbers.

a. $700 + 9 =$ _____	**b.** $70 + 600 + 4 =$ _____
c. $80 + 500 =$ _____	**d.** $8 + 500 + 50 =$ _____

4. Count by fives: _____ , _____ , _____ , _____ , 715, 720.

5. Write the numbers that are 10 less and 10 more than the given number.

    a. _____ , 292, _____

    b. _____ , 545, _____

6. Count by 20s, and fill in the grid.

200	220	240		
300				

7. Compare. Write < or > in the box.

a. 238 ☐ 265	b. 391 ☐ 193	c. 405 ☐ 450	d. 981 ☐ 819
e. 8 + 600 ☐ 60 + 800		f. 30 + 300 + 5 ☐ 90 + 8 + 100	

8. Arrange the three numbers in order, from the smallest to the biggest.

a. 109, 901, 199	b. 717, 175, 177

9. Add in columns.

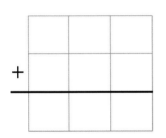
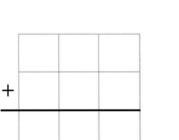

10. Add and subtract whole hundreds.

a.	b.	c.
720 + 200 = _____	508 + 400 = _____	219 + 500 = _____
780 − 300 = _____	670 − 400 = _____	954 − 900 = _____

11. Add and subtract whole tens. You can underline the tens to help you.

a.	b.	c.
5<u>8</u>0 + <u>2</u>0 = _____	969 − 40 = _____	572 − 30 = _____
620 + 70 = _____	433 + 20 = _____	884 − 70 = _____

12. Fill in.

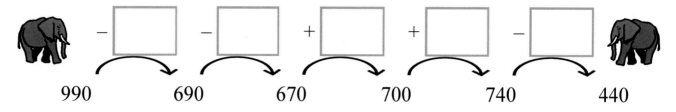

990   690   670   700   740   440

13. Solve the word problems.

a. Grandpa had 250 sheep. During 1 week, wolves killed 10 of them,
but 20 new little lambs were born.
How many sheep were in the flock at the end of the week?

b. Jake has 170 fish in his aquariums. Of all the fish, 30 are rainbow fish,
and the rest are goldfish. How many are goldfish?

c. Jake bought 50 more goldfish and 70 more rainbow fish.
How many goldfish does he have now?

And how many rainbow fish?

d. Sandra traveled 400 km in an airplane, and then 30 km in a car,
to go visit her mother. Then she returned the same way.
How many kilometers did Sandra travel?

	**a.** If you count by 10s from this number 3 times, you will get to 62.
	b. It is less than 10. If you double it, you get a number that is more than 10, but it will not be 14, 18, or 12.

Three-Digit Numbers Test

1. **a.** Count by fives:

475, 480, _____ , _____ , _____ , _____ , _____

   b. Count by tens:

376, 386, _____ , _____ , _____ , _____ , _____

2. Break these numbers into hundreds, tens, and ones.

a.	b.
235 = _____ + _____ + _____	805 = _____ + _____ + _____

3. These numbers are broken into their hundreds, tens, and ones. Write the numbers.

a. $600 + 80 + 8 =$ _____	**b.** $80 + 200 + 5 =$ _____
$400 + 60 =$ _____	$100 + 6 =$ _____

4. Write either < or > between the numbers.

a. 159 300	**b.** 323 230	**c.** 450 504	**d.** 482 284

5. Arrange the numbers in order.

a. 689, 869, 986	**b.** 524, 245, 452
_____ < _____ < _____	_____ < _____ < _____

6. Compare the expressions and write <, >, or = .

   a. $6 + 200 + 50$ ☐ 256 **b.** $800 + 9$ ☐ $90 + 800$

   c. $400 + 60 + 2$ ☐ $40 + 6 + 200$ **d.** $3 + 700$ ☐ $700 + 6$

7. One of the "parts" for the numbers is missing. Find out what number the triangle means.

a. $300 +$ △ $+ 7 = 347$ △ $=$ _____	**b.** $900 +$ △ $+ 40 = 948$ △ $=$ _____	**c.** $5 +$ △ $+ 80 = 585$ △ $=$ _____

8. Add and subtract.

a.	b.	c.
$765 - 200 =$ _____	$802 - 400 =$ _____	$778 - 500 =$ _____
$548 - 300 =$ _____	$980 - 600 =$ _____	$994 - 900 =$ _____

9. Children counted cars that were passing by while waiting for the bus.

One in the pictograph means 5 cars.

Cars that children counted	
Jayden	
Natalie	
Caleb	

= 5 cars

a. How many cars did Natalie count?

b. How many did Jayden count?

c. How many more did Natalie count than Caleb?

Mixed Review 8

1. Write the time that the clock shows, and the time 5 minutes later.

	a.	b.	c.	d.
	_____ : _____	_____ : _____	_____ : _____	_____ : _____
5 min. later →	_____ : _____	_____ : _____	_____ : _____	_____ : _____

2. Find the missing numbers.

a. $16 - \boxed{} = 9$	**b.** $11 - \boxed{} = 3$	**c.** $12 - \boxed{} = 9$
d. $\boxed{} - 8 = 6$	**e.** $\boxed{} - 7 = 5$	**f.** $\boxed{} - 9 = 9$

3. Add. Compare the problems.

a. $7 + 6 =$ _____	**b.** $8 + 9 =$ _____	**c.** $5 + 8 =$ _____
$17 + 6 =$ _____	$68 + 9 =$ _____	$35 + 8 =$ _____

4. Mom divided 16 plums evenly between Jane and John.
   John ate 3 of his. Jane ate 2 of hers.

   How many does John have left?

   How many does Jane have left?

5. Write each number as a double of some other number.

a. $12 =$ ____ + ____	**b.** $18 =$ ____ + ____	**c.** $100 =$ _____ + _____
d. $40 =$ ____ + ____	**e.** $14 =$ ____ + ____	**f.** $600 =$ _____ + _____

6. Add.

a.	b.	c.	d.	e.
7 5	1 8	2 4	3 7	5 1
2 6	2 7	5 5	2 8	2 9
+ 2 4	+ 5 9	+ 2 5	3 7	9
			+ 2 3	+ 1 5

7. Draw a square in the grid that has 9 little squares inside.

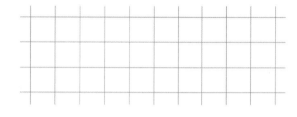

8. Draw a rectangle in the grid that has 12 little squares inside. Can you draw another one with a different shape?

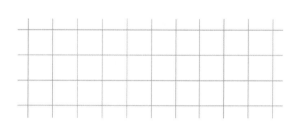

9. Divide the shapes into two, three, or four equal parts so that you can color the fraction.

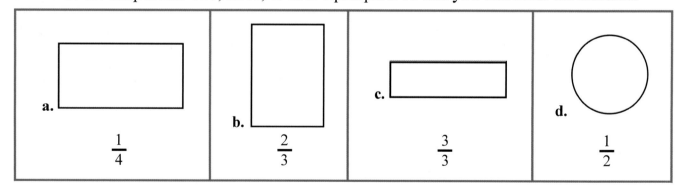

a. $\frac{1}{4}$ b. $\frac{2}{3}$ c. $\frac{3}{3}$ d. $\frac{1}{2}$

10. Color. Then compare and write < , > , or = . Which is more "pie" to eat?

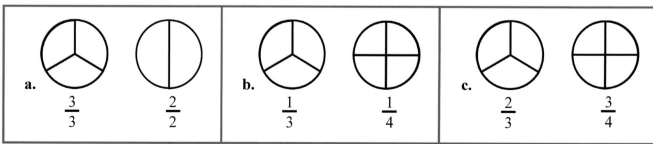

a. $\frac{3}{3}$ $\frac{2}{2}$ b. $\frac{1}{3}$ $\frac{1}{4}$ c. $\frac{2}{3}$ $\frac{3}{4}$

Mixed Review 9

1. Write what place the teddy bear has using ordinal numbers.

    a. The _____ place from the left.

    b. The _____ place from the right.

    c. The _____ place from the left.

    d. The _____ place from the right.

2. Jack is on the track team. He spends a half-hour for warm-up
   exercises, an hour running, and thirty minutes jumping hurdles.
   How much time does he spend practicing?

3. One T-shirt costs $12 and another cost $5 more than that.
   If you buy both, what is the total cost?

4. Add.

a.	b.	c.	d.	e.
3 7	2 9	5 4	3 6	2 8
1 8	8 0	1 3	9	1 8
+ 4 3	+ 3 6	+ 7 6	+ 4 3	+ 3 6

5. Complete the **next whole ten**.

a.	b.	c.
66 + ____ = 70	31 + 3 + ____ = 40	47 + ____ + 1 = 50
92 + ____ = _____	63 + 2 + ____ = 70	32 + ____ + 2 = 40

6. Write the time with *hours:minutes*, and using "past" or "till."

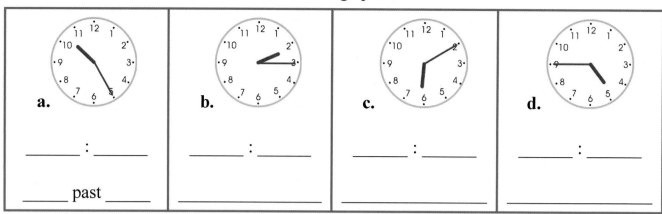

a. _____ : _____

_____ past _____

b. _____ : _____

_____

c. _____ : _____

_____

d. _____ : _____

_____

7. Draw here six dots randomly
and join them like a dot-to-dot.
Use a ruler. What shape do
you get?
(Hint: It will not be a straight line.)

8. Color <u>one whole pie</u>.
Write <u>one</u> as a fraction,
in many different ways.

a. 1 =   **b.** 1 = _____   **c.** 1 = _____

9. Divide these shapes. Then color as you are asked to.

a.

b.

c.

d.

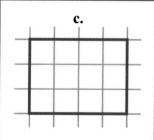

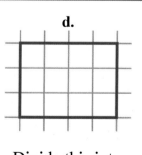

Divide this into
halves. Color $\frac{1}{2}$.

Divide this into
fourths. Color $\frac{3}{4}$.

Divide this into
thirds. Color $\frac{1}{3}$.

Divide this into
fourths. Color $\frac{1}{4}$

Measuring Review

1. Which unit or units would you use for the following distances: inches (in.), feet (ft), miles (mi), centimeters (cm), or meters (m)? If two different units work, write both.

Distance	Unit or units
from your house to the grocery store	
from Miami to New York	
the distance across the room	
the height of a bookcase	

2. Measure this line to the nearest centimeter and to the nearest half-inch.

   ████████████████ about _____ cm _or_ about _____ in

3. **a.** Draw a line that is
      3 1/2 inches long.

   b. Draw a line that is
      9 cm long.

4. Measure these two pencils to the nearest centimeter, *and* to the nearest half-inch. Then fill in:

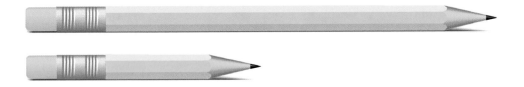

   The longer pencil is about _____ cm longer than the shorter one.

   The longer pencil is about _____ inches longer than the shorter one.

5. Measure the width and length of the room you are in. First, measure them using feet and inches. Then, measure them using meters and centimeters.

   Width: _____ ft _____ in _or_ _____ m _____ cm

   Length: _____ ft _____ in _or_ _____ m _____ cm

Measuring Test

1. Cross out the sentences that don't make sense.

   a. An 11-year old boy weighs 12 pounds.   **b.** An elephant is 3 m tall.

   c. My science book is 25 m wide.   **d.** The suitcase weighs 400 kg.

2. Measure these pencils two times, to the nearest half-inch, and to the nearest centimeter.

Pencil #1	in.	cm
Pencil #2	in.	cm

3. **a.** Draw a line that is
       4 1/2 inches long.

   b. Draw a line that is
       9 cm long.

4. Arrange these measuring units from the shortest to the longest.

   kilometer   inch   centimeter   foot

5. Andy measured how long a piece of rope is. It was 10 feet long. Then he measured
   the same rope in meters. Which measurement did Andy get?   **10 m   3 m   30 m**

6. Choose a unit to measure these: centimeters (cm), meters (m), or kilometers (km).

Distance	Unit		Distance	Unit
from Florida to California			length of a garden	
around your head			height of a room	

7. The teacher needs to arrange this task beforehand, and check students' results.

   Your teacher gives you an item. Find out how heavy it is. _____

Mixed Review 10

1. Fill in.

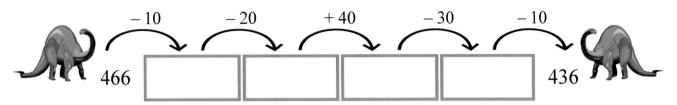

466 [ ] -10 [ ] -20 [ ] $+40$ [ ] -30 [ ] -10 436

2. One of the "parts" for the numbers is missing. Solve what number the triangle means.

a. $700 + \triangle + 5 = 735$	**b.** $400 + 40 + \triangle = 449$	**c.** $7 + \triangle + 90 = 297$
$\triangle =$ _____	$\triangle =$ _____	$\triangle =$ _____

3. Skip-count by tens.

a. 806, 816, _____, _____, _____, _____, _____, _____

b. 542, 532, _____, _____, _____, _____, _____, _____

4. Skip-count by fives.

a. 280, 285, _____, _____, _____, _____, _____, _____

b. 1000, 995, _____, _____, _____, _____, _____, _____

5. Find the double of these numbers.

    a. Double 150   **b.** Double 247   **c.** Double 499

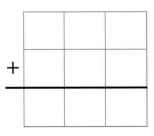 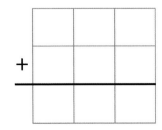

6. Draw some lines on a blank paper. Use a <u>ruler</u>. Hold the ruler down tight with one hand, while drawing the line with the other. It takes some practice!

   a. 3 in long

   b. 8 in long

   c. 12 cm long

   d. 7 cm long

7. Aaron visited an amusement park with his family that was 90 km away. They drove there and back. How many kilometers did the family drive all totaled?

8. Which unit would you use to find these below: centimeters (cm), meters (m), or kilometers (km)?

Distance	Unit
from Pretoria to Durban	
the length of your room	
the length of a pencil	

Distance	Unit
around your neck	
the width of a butterfly	
how far you can throw a ball	

9. The total is missing from the subtraction sentence. Solve.

a. ☐ − 5 = 24   **b.** ☐ − 60 = 420   **c.** ☐ − 500 = 240

10. Write the time using the **hours:minutes** way. Use your practice clock to help.

a. 10 past 8	b. 15 till 7	c. 25 past 12	d. half-past 7
____ : ____	____ : ____	____ : ____	____ : ____
e. 9 o'clock	f. 20 till 6	g. 5 till 11	h. 25 till 4
____ : ____	____ : ____	____ : ____	____ : ____

11. Write the dates in the form (*month*) (*day*) (*year*), such as June 15, 2016.

	month	day	year
a. today's date	_____	_____	_____
b. tomorrow's date	_____	_____	_____
c. your birthday this year	_____	_____	_____
d. your friend's birthday this year	_____	_____	_____

12. **a.** Draw here a big and a small three-sided shape. What are three-sided shapes called?

b. Draw here a red six-sided shape and a blue four-sided shape.
What are six-sided shapes called?

The DIGITS of the number 467 are 4, 6, and 7. The sum of its digits is $4 + 6 + 7 = 17$ (just add its digits).

Find a number that...

- is more than 100 but less than 200;
- the sum of its digits is 11.

There are actually 9 different numbers like that.
Can you find all of them?

Puzzle Corner

Mixed Review 11

1. Solve.

a. Diane needs 8 apples to make one pie, and she wants to make *two* pies for a bake sale. Diane already has 10 apples. How many more apples does Diane need to buy?

b. Joe and Ben are building go-carts. They sold one for $18. How much would two go-carts cost?

c. Raylene has four cats. One of them had kittens. Now she has double as many cats as before. How many cats does Raylene have now?

How many of them are kittens?

2. Draw the hands on the clock faces to show the given time.

  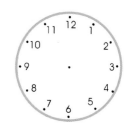

    a. 1:25   **b.** 3:15   **c.** 15 till 3

3. Find the pattern and continue it.

a.

577 − 10 = _____

577 − 20 = _____

577 − 30 = _____

577 − ____ = _____

_____ − ____ = _____

_____ − ____ = _____

b.

926 − 0 = _____

926 − 100 = _____

926 − 200 = _____

_____ − _____ = _____

_____ − _____ = _____

_____ − _____ = _____

4. Solve.

a. 9 + 8 = _____	**b.** 8 + 8 = _____	**c.** 5 + 8 = _____
5 + 6 = _____	6 + 6 = _____	4 + 7 = _____
7 + 7 = _____	9 + 7 = _____	6 + 8 = _____

5. Barbara drew some shapes in her notebook.
   She needs you to help her label them.

  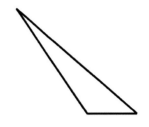

a. _____   **b.** _____   **c.** _____

6. In the grid, draw a rectangle that is
   6 units wide and 3 units long.

   How many square are inside it?

   _____ squares

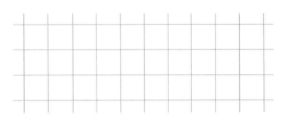

7. Fill in the missing numbers and words in the charts below.

Ordinal Number	Name
	first
2nd	
	fourth
	seventh

Ordinal Number	Name
8th	
	ninth
10th	
	thirteenth

Regrouping in Addition and Subtraction Review

1. Add.

a.
```
   2 1 5
 + 4 7 7
 _____
```

b.
```
   1 9 2
 + 2 2 5
 _____
```

c.
```
   3 0 3
   1 2 8
 + 2 8 7
 _____
```

d.
```
   4 0 9
   2 1 9
 + 1 3 6
 _____
```

2. Sarah bought three bicycles for her children.
   Each bicycle cost $154.
   How much was the total cost?

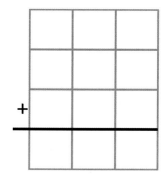

3. Add mentally. THINK of the new hundred you might get
   from adding the tens.

a.	b.	c.
80 + 40 = _____	90 + 90 = _____	690 + 50 = _____
780 + 40 = _____	240 + 50 = _____	470 + 80 = _____

4. Find how many feet it is if you walk all
   of the way around this rectangle.

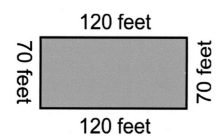

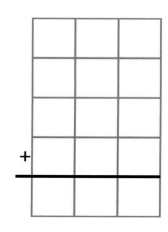

5. Subtract. Regroup if necessary. Check each subtraction by
   adding your answer and the number you subtracted.

a.		b.	
$\begin{array}{r} 8\ 8 \\ -\ 5\ 4 \\ \hline \end{array}$ $\quad \begin{array}{r} \\ +\ 5\ 4 \\ \hline \end{array}$		$\begin{array}{r} 6\ 3 \\ -\ 4\ 8 \\ \hline \end{array}$ $\quad \begin{array}{r} \\ + \\ \hline \end{array}$	
c.		d.	
$\begin{array}{r} 8\ 4 \\ -\ 4\ 9 \\ \hline \end{array}$ $\quad \begin{array}{r} \\ + \\ \hline \end{array}$		$\begin{array}{r} 8\ 8\ 2 \\ -\ 1\ 5\ 9 \\ \hline \end{array}$ $\quad \begin{array}{r} \\ + \\ \hline \end{array}$	
e.		f.	
$\begin{array}{r} 5\ 5\ 6 \\ -\ 3\ 9\ 1 \\ \hline \end{array}$ $\quad \begin{array}{r} \\ + \\ \hline \end{array}$		$\begin{array}{r} 5\ 5\ 0 \\ -\ 2\ 4\ 6 \\ \hline \end{array}$ $\quad \begin{array}{r} \\ + \\ \hline \end{array}$	

6. Subtract using mental math methods.

a. $15 - 7 = $ _____ $55 - 7 = $ _____	b. $13 - 5 = $ _____ $93 - 5 = $ _____	c. $82 - 77 = $ _____ $45 - 41 = $ _____
d. $80 - 71 = $ _____ $100 - 95 = $ _____	e. $56 - 40 = $ _____ $56 - 43 = $ _____	f. $78 - 35 = $ _____ $33 - 4 = $ _____

7. Find what numbers are missing.

a.
$$\begin{array}{r} 2\ \square\ 4 \\ +\ 4\ 7\ 7 \\ \hline 7\ 3\ 1 \end{array}$$

b.
$$\begin{array}{r} 5\ \square\ 9 \\ +\ \square\ 2\ 5 \\ \hline 9\ 1\ 4 \end{array}$$

c.
$$\begin{array}{r} 2\ 0\ \square \\ +\ 6\ \square\ 6 \\ \hline 8\ 9\ 2 \end{array}$$

d.
$$\begin{array}{r} 6\ 8\ \square \\ +\ \square\ 1\ 9 \\ \hline 9\ 0\ 0 \end{array}$$

8. Solve.

a. Some people are riding on the bus. At the bus stop, 13 people get on. Now there are 52 people on the bus. How many were there originally?

b. Molly has 23 stuffed toys that she likes, and 16 that she does not like.

How many stuffed toys does Molly have?

c. Molly gave the 16 toys she does not like to her sister Annie. Now, Annie has 33 toys.

How many toys did Annie have before?

d. Jessica had 465 points in a computer game.
She played and got 145 more points.
Then she also got a 90-point bonus!
How many points does Jessica have now?

e. Olivia did 26 jumping jacks, which was 14 fewer jumping jacks than what her brother Aaron did. How many jumping jacks did Aaron do?

9. **a.** Fill in the table with how many points the children got in the game.

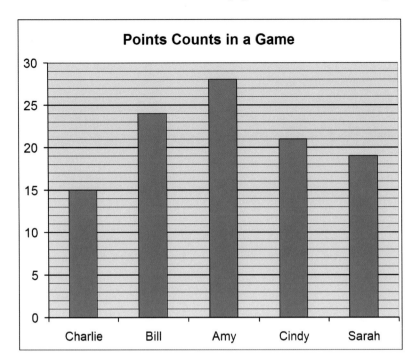

CHILD	POINTS
Charlie	15
Bill	
Amy	
Cindy	
Sarah	

b. How many fewer points did Bill get than Amy?

c. How many more points did Cindy get than Charlie?

Can you place numbers from 1 through 12 into the circles so that the sum of each connecting line is 26?

Hint: The numbers that go in the top corners are 7 and 6, and the numbers that go in the bottom corners are 5 and 8.

Puzzle Corner

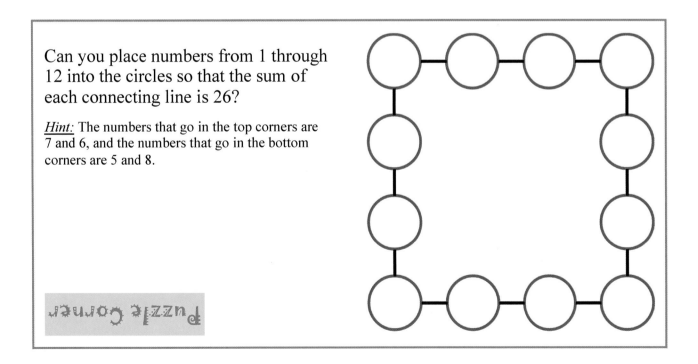

Regrouping in Addition and Subtraction Test

1. Add and subtract.

a.	b.	c.	d.
$\begin{array}{r} 2\ 1\ 9 \\ +\ 4\ 3\ 5 \\ \hline \end{array}$	$\begin{array}{r} 5\ 6\ 2 \\ +\ 3\ 7\ 5 \\ \hline \end{array}$	$\begin{array}{r} 4\ 9\ 6 \\ +\ 2\ 8\ 6 \\ \hline \end{array}$	$\begin{array}{r} 6\ 2 \\ -\ 2\ 7 \\ \hline \end{array}$

2. Subtract. Check by adding the result and what was subtracted.

a.		b.	
$\begin{array}{r} 9\ 6\ 4 \\ -\ 2\ 2\ 7 \\ \hline \end{array}$	+ _____	$\begin{array}{r} 7\ 4\ 8 \\ -\ 3\ 7\ 2 \\ \hline \end{array}$	+ _____

3. Add and subtract mentally.

a.	b.	c.
80 + 40 = _____	690 + 60 = _____	93 – 52 = _____
280 + 50 = _____	85 – 31 = _____	91 – 89 = _____

4. Find the total bill when Nancy paid for three nights in a hotel, at $129 per night.

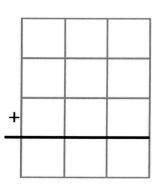

5. Solve the word problems.

a. There were 250 sacks of wheat in a storehouse. Then the store owner sold 68 sacks.

How many sacks are left?

b. A pet store has 52 kittens. Of them, 15 are white and 18 are orange. The rest are black.

How many are black?

c. Jackie bought two 15-lb bags of cat food and one 50-lb bag of dog food. What is the total weight of these bags?

d. A store sold 47 coffee makers in January. In February, the store sold 19 fewer coffee makers than in January. How many coffee makers did the store sell in February?

How many coffee makers did the store sell in those two months?

e. Grandpa walked 300 meters on Tuesday. The next day he walked 120 meters more than on Tuesday. How many meters did he walk in those two days *in total*?

57

Mixed Review 12

1. Under each number in the chart, write its DOUBLE. Notice what pattern it makes!

8	9	10	11	12	13	14	15	16	17

2. Solve. Use the chart above.

   a. Two girls shared evenly 30 marbles. How many did each get?

   b. A bag of potatoes weighs 28 kg. The family ate half of it.
   How many kilograms of potatoes are left?

   c. Katy had $60. She spent half of it to buy a gift.
   Then, Katy bought a toy for $9.
   How many dollars does Katy have now?

   d. Mom used up half of the apples she had to make a pie.
   Now she has 8 apples. How many did she have to start with?

3. Subtract whole hundreds and whole tens.

a.	b.	c.
$239 - 100 =$ _____	$871 - 400 =$ _____	$704 - 500 =$ _____
d.	**e.**	**f.**
$376 - 40 =$ _____	$781 - 20 =$ _____	$1000 - 50 =$ _____

4. **a.** Circle all the months in this list that have 31 days.

January February March April May June July August September October November December

   b. Circle all the months in this list that have 30 days.

January February March April May June July August September October November December

   c. Which month didn't get circled either time? _____

   How many days does it have? _____

5. Color. Then compare and write < , > , or = . Which is more "pie" to eat?

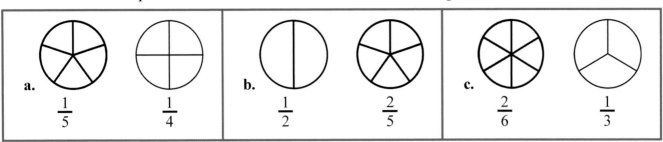

6. Find out what number the triangle means. Also explain how you do it!

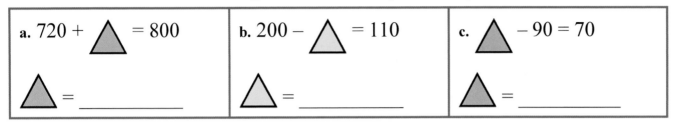

7. Write the time using the wordings "past" or "till", and using numbers.

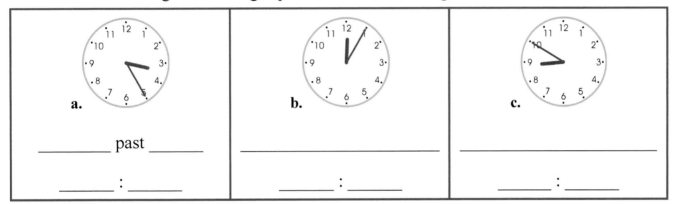

8. Measure many different erasers to the nearest whole centimeter. Then make a line plot.
   This means that you mark an "x" for each eraser above the number line.

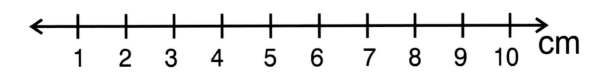

Mixed Review 13

1. Cross out and subtract. Subtract also in columns!

a. 40 – 29 = _____

b. 50 – 28 = _____

c. 62 – 25

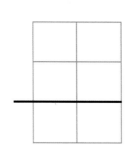

d. 83 – 46

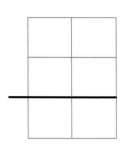

2. Add.

a. 6 + 3 = _____	b. 7 + 5 = _____	c. 8 + 6 = _____	d. 9 + 9 = _____
6 + 10 = _____	7 + 8 = _____	8 + 7 = _____	9 + 4 = _____

3. Find these differences. Think of adding more.

a. 17 – 11 = _____	b. 43 – 37 = _____	c. 66 – 59 = _____
Think: 11 + _____ = 17	Think: 37 + _____ = 43	Think: 59 + _____ = 66
d. 35 – 28 = _____	e. 80 – 77 = _____	f. 100 – 94 = _____

4. Find what was subtracted.

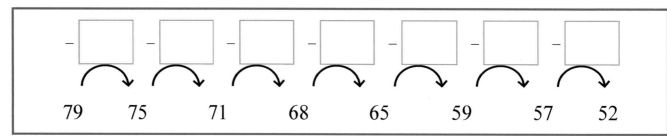

5. Write the times using hours : minutes.

a. 15 past 6	b. 20 till 3	c. 5 past 10	d. half past 3
_____ : _____	_____ : _____	_____ : _____	_____ : _____
e. 15 till 8	f. 20 till 12	g. 5 till 1	h. 25 past 1
_____ : _____	_____ : _____	_____ : _____	_____ : _____

6. Dan weighs 138 pounds. His sister weighs twenty pounds less than he does.

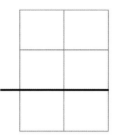

   a. How much does Dan's sister weigh?

   b. How much do they weigh together?

7. a. Draw a rectangle. Use a ruler to make it as neat as you can!

   b. Draw a line through the rectangle from one corner to the opposite corner.

   c. What shapes are formed now?

8. a. In each picture, color TWO slices of the whole pie, and write the fraction.

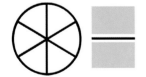 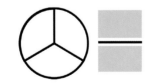 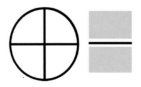 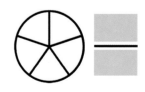

   b. Now, find the largest fraction (the one that has most to eat).

9. Compare, and write < or > .

   a. 106 ☐ 120   b. 141 ☐ 114   c. 700 + 80 + 9 ☐ 90 + 8 + 700

Money Review

1. How much is the total if you have:

a. a quarter, a nickel, and three pennies	**b.** three dimes and four nickels
_____¢	_____¢

2. Make these money amounts. Use real money or draw. Use at least one quarter.

a. 28¢	**b.** 93¢

3. Write the amount in *dollars*.

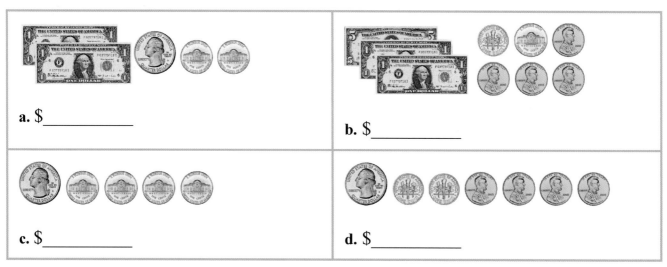

a. $_____

b. $_____

c. $_____

d. $_____

4. Write how many cents you give and how many cents is your change.

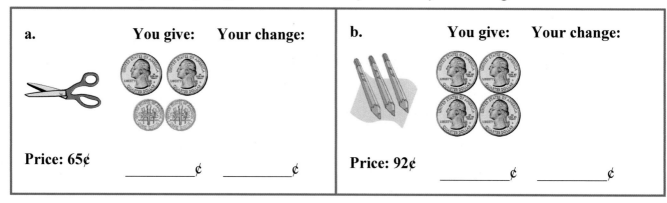

a.	You give:	Your change:
Price: 65¢	_____¢	_____¢

b.	You give:	Your change:
Price: 92¢	_____¢	_____¢

5. Count up to find the change. Draw the coins for the change.

a.  $2.15

Customer gives $3. Change: _____

b. $1.59

Customer gives $2. Change: _____

c. $4.85

Customer gives $5. Change: _____

6. Lily has $1.26. Alex has two dimes, two quarters,
   and seven pennies in his piggy bank.
   How much money does Alex have?

   How much money do the children have together?

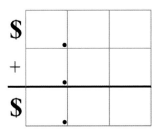

7. Find the total cost of buying the things listed.

$0.98

$1.65

$0.78

a. a yogurt and an apple

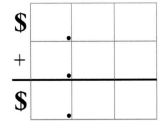

b. two yogurts and a sandwich

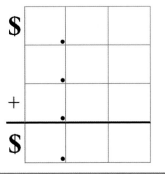

Money Test

1. How much money? Write the amount.

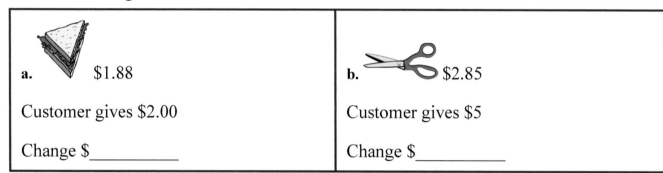

a. $_____

b. $_____

c. $_____

d. $_____

2. Find the change.

a. $1.88	**b.** $2.85
Customer gives $2.00	Customer gives $5
Change $_____	Change $_____

3. Find the total cost.

a. Matt bought two sandwiches for $1.56 each and water for $0.78.

b. Eva bought two sets of water paints for $2.55 each.

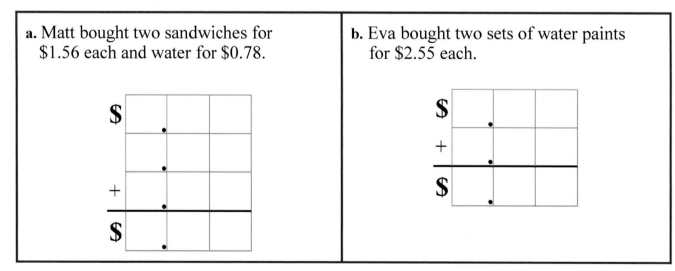

Mixed Review 14

1. Write the numbers in columns and add.

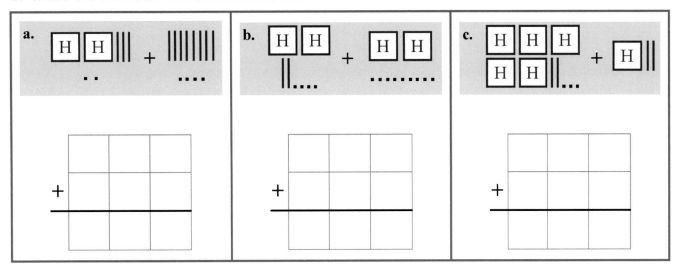

2. Solve.

a. Mary has 78 marbles. Her brother had 13 fewer marbles than her. How many marbles do the children have together?

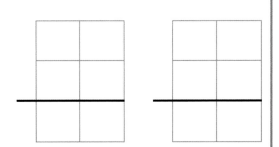

b. Mom had $250 with her when she went shopping. She bought groceries for $120, and gasoline for $50. How much money does she have left now?

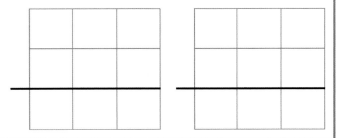

c. Pablo has read 141 pages of a book that has 213 pages. How many pages does he have left to read?

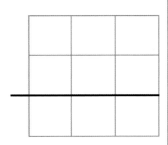

3. Subtract. Regroup if necessary. Check each subtraction by *adding your answer and the number you subtracted.*

a.		b.	
$\begin{array}{r} 9\ 3 \\ -\ 2\ 8 \\ \hline \end{array}$ $\begin{array}{r} \\ +\ 2\ 8 \\ \hline \end{array}$		$\begin{array}{r} 5\ 2\ 8 \\ -\ 2\ 4\ 5 \\ \hline \end{array}$ $+$	

4. Measure the sides of this triangle BOTH in inches, to the nearest half-inch, and in centimeters, to the nearest centimeter. Write your results in the table.

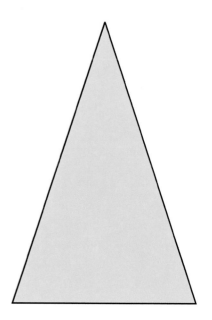

Triangle	in inches	in centimeters
Side 1	in	cm
Side 2	in	cm
Side 3	in	cm

5. The pictograph shows how many fish the family members caught when they went fishing.

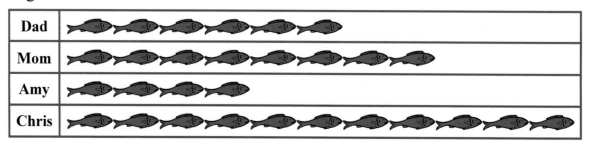

a. Who caught the most fish?

b. How many more fish did Chris catch than Dad?

c. How many fewer fish did Amy catch than Mom?

d. How many did Amy and Chris catch together?

6. The bar graph shows how many toy cars some kids have. Chloe has 14 cars.
   Draw a bar for her in the graph.

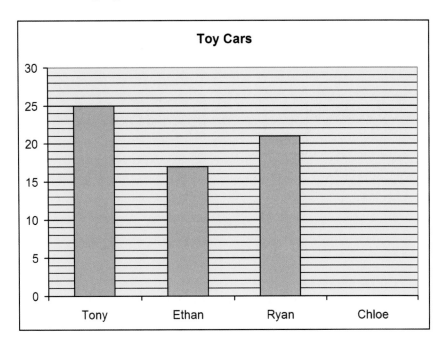

a. How many cars does Ethan have?

b. How many more cars does Tony have than Ethan?

c. How many cars do Ryan and Chloe have together?

d. If Ethan gives Ryan 5 cars, will Ryan then have more than Tony?

Puzzle Corner

Can you place numbers from 1 through 9 into the circles so that the sum of each side of the triangle is 20?

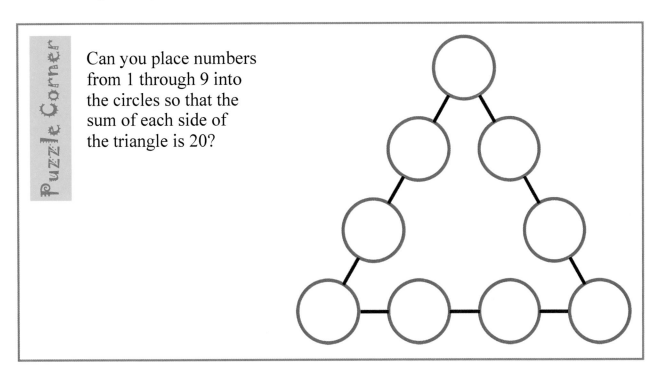

Mixed Review 15

1. Color the part indicated.

a. $\dfrac{1}{2}$ b. $\dfrac{1}{4}$ c. $\dfrac{3}{4}$ d. $\dfrac{2}{4}$ e. $\dfrac{4}{4}$ f. $\dfrac{2}{2}$

2. How many hours pass?

a. From 3:00 to 8:00 _____ hours	**d.** From noon till 4 PM _____ hours
b. From 6 AM to 1 PM _____ hours	**e.** From 7 PM to 11 PM _____ hours
c. From 7 PM to midnight _____ hours	**f.** From 10 AM to 2 PM _____ hours

3. Add and subtract.

a. $7 + 8 =$ _____	**b.** $8 +$ _____ $= 13$	**c.** $14 -$ _____ $= 7$
$4 + 9 =$ _____	$8 -$ _____ $= 2$	$6 +$ _____ $= 11$
$15 - 9 =$ _____	$4 +$ _____ $= 11$	$19 -$ _____ $= 12$

4. Add money amounts in columns.

    a. $\$0.39 + \$0.55$ **b.** $\$2.25 + \$0.89$ **c.** $\$5.53 + \$2.69$

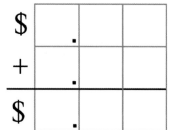 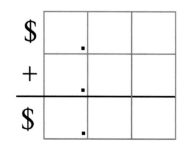 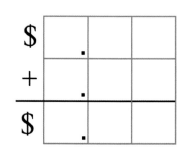

5. **a.** How many corners does the shape have? _____

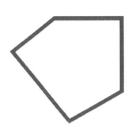

   b. What is the shape called? _____

6. In a game, Amy has 15 marbles and John has 5 fewer marbles than Amy.
   How many does John have?

   How many marbles do Amy and John have together?

7. Jeremy ate 4 slices of pie, which was two fewer pieces than what Eva ate.
   How many did Eva eat?

8. Add and subtract mentally.

a.	b.	c.
$507 + 30 =$ _____	$640 - 40 =$ _____	$552 - 20 =$ _____
$507 + 300 =$ _____	$640 - 400 =$ _____	$552 - 200 =$ _____

9. Subtract. Check by adding!

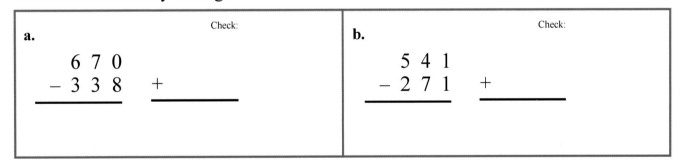

10. Count up to find the change. You can draw in the coins to help you.

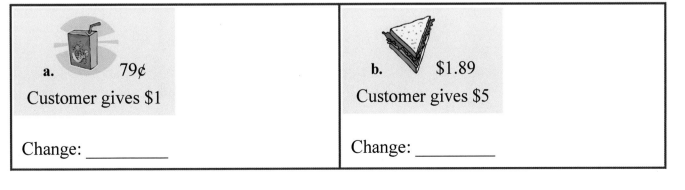

69

Exploring Multiplication Review

1. Draw groups to illustrate the multiplications.

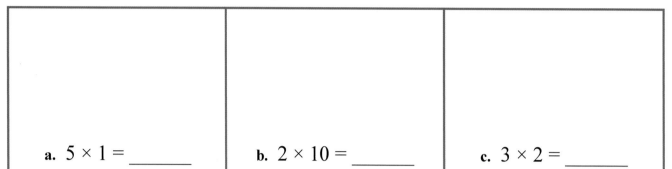

    a. $5 \times 1 =$ _____   **b.** $2 \times 10 =$ _____   **c.** $3 \times 2 =$ _____

2. Draw number-line jumps to illustrate these multiplication sentences.

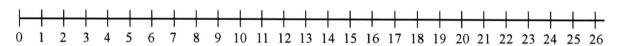

    a. $3 \times 6 =$ _____

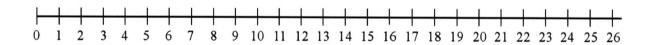

    b. $4 \times 3 =$ _____

3. Write each multiplication as an addition.

a. 3×3	**b.** 4×2

4. Multiply.

a. $2 \times 3 =$ _____	**b.** $3 \times 3 =$ _____	**c.** $1 \times 3 =$ _____
$2 \times 10 =$ _____	$3 \times 20 =$ _____	$2 \times 0 =$ _____
$2 \times 20 =$ _____	$7 \times 1 =$ _____	$4 \times 5 =$ _____

5. Hannah made many math problems from the same picture. Solve the problems.

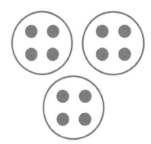

   a. $4 + 4 + 4 =$ _____   **b.** $12 - 4 =$ _____

   c. $4 + 4 =$ _____   **d.** $12 - 4 - 4 =$ _____

   e. $3 \times 4 =$ _____   **f.** $1 \times 4 =$ _____

6. Make as many math problems as you can from this picture:

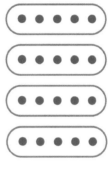

7. Solve.

   a. Katie had 5 vases. She put ten flowers in each vase.
   What is the total number of flowers in all the vases?

   b. John had 20 toy cars and Jim had 10. John gave half of his to Jim.
   Now who has more cars?

   How many more?

8. Draw a line from the problems to 10 or 50 if they are equal to 10 or 50.

$5 + 6$	$29 - 9$	2×5	$25 + 25$	$61 - 11$	5×10
$15 - 5$		$16 - 3 - 3$	$90 - 50$		$45 + 3 + 5$
$\frac{1}{2}$ of 10	**10**	0×10	$\frac{1}{2}$ of 100	**50**	1×50
		$\frac{1}{2}$ of 20			$\frac{1}{2}$ of 80
5×2			10×10		
$3 + 3 + 3$	1×5	$6 + 4$	$20 + 20 + 10$	2×20	$70 - 20$

Exploring Multiplication Test

1. Draw groups to illustrate the multiplication.

a. $6 \times 1 =$ _____	**b.** $2 \times 7 =$ _____	**c.** $3 \times 3 =$ _____

2. Write each addition as a multiplication.

a. $6 + 6 + 6 + 6 =$ _____ $\times$ _____

b. $50 + 50 + 50 =$ _____ $\times$ _____

3. Write each multiplication as an addition.

a. $2 \times 8 =$ _____

b. $5 \times 3 =$ _____

4. Draw number-line jumps for these multiplications.

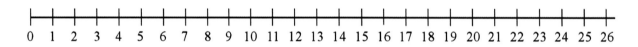

a. $6 \times 3 =$ _____

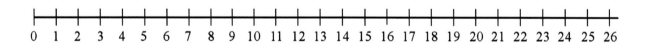

b. $5 \times 5 =$ _____

5. Multiply.

a. $4 \times 3 =$ _____ $3 \times 10 =$ _____	**b.** $5 \times 0 =$ _____ $2 \times 20 =$ _____	**c.** $1 \times 6 =$ _____ $2 \times 9 =$ _____
d. $4 \times 3 =$ _____ $2 \times 8 =$ _____	**e.** $2 \times 12 =$ _____ $3 \times 5 =$ _____	**f.** $4 \times 10 =$ _____ $1 \times 800 =$ _____

Mixed Review 16

1. Subtract in your head.

a. 644 – 20 = _____	**b.** 777 – 70 = _____	**c.** 98 – 26 = _____
644 – 400 = _____	777 – 500 = _____	100 – 96 = _____

2. How much money? Write the amount.

a. $_____	**b.** $_____

3. Amy's piggy bank had 57¢. Brett's piggy bank had four quarters.

   How many _cents_ does Brett have?

   How many cents do the two children have
   if they put their money together?

4. You buy something. Find the total. Then find the change, and draw the coins that
   would be your change.

a. A tennis ball costs $1.10. You give $5.
Total cost: _____
Change: _____
b. A drink costs 80¢. You buy two of them. You give $2.
Total cost: _____
Change: _____

5. Write the time using the wordings "past" or "till." Also write the time using numbers.

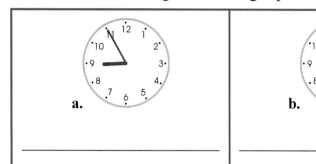

a. _____

_____ : _____

b. _____

_____ : _____

c. _____

_____ : _____

6. Write the time 5 minutes later than what the clocks show in the previous exercise.

a. _____ : _____

b. _____ : _____

c. _____ : _____

7. Subtract. Regroup if necessary. Check each subtraction by *adding your answer and the number you subtracted.*

a.
```
    6 5 2
  − 2 2 7          +
  _____      _____
```

b.
```
    5 4 8
  − 1 8 5          +
  _____      _____
```

8. Find how many meters it is if you walk all the way around this park.

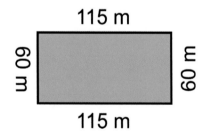

115 m

60 m

115 m

60 m

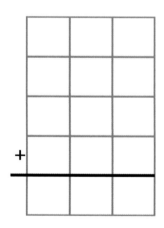

+

9. Find out what number the triangle means.

a. 560 +  = 600

△ = _____

b. 400 − △ = 310

△ = _____

c. △ − 60 = 60

△ = _____

74

10. Draw rectangles so they have a certain number of little squares inside. Guess and check!

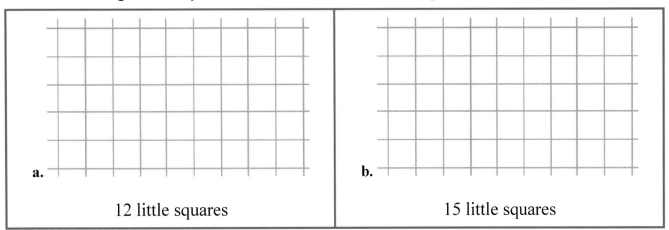

a.

12 little squares

b.

15 little squares

11. A *face* is any of the flat sides of a solid.

a. Count how many faces a cube has. _____ faces

What shapes are they? _____

b. Count how many faces a box has. _____ faces

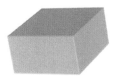

What shapes are they? _____

12. What shapes are these?

a.

b.

_____

_____

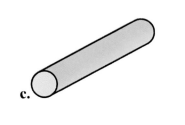

c.

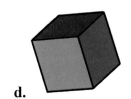

d.

_____

_____

Mixed Review 17

1. Color in the chart all the even numbers.

1	2	3	4	5	6	7	8	9	10
11	12	13	14	15	16	17	18	19	20
21	22	23	24	25	26	27	28	29	30

2. **a.** Today is January 5. I am going away for three weeks and two days. What day will I return? (See the calendar on the right.)

January

Su	Mo	Tu	We	Th	Fr	Sa
		1	2	3	4	5
6	7	8	9	10	11	12
13	14	15	16	17	18	19
20	21	22	23	24	25	26
27	28	29	30	31		

   b. I went to the gym every Wednesday in January. What were the dates I went to the gym?

3. Aunt Cindy gave Aiden and Samantha $30. The children shared the money equally. Samantha already had $5 in her piggy bank. How much money does Samantha have now?

4. Subtract.

a.
$$975 - 246$$

b.
$$629 - 189$$

c.
$$514 - 323$$

d.
$$650 - 126$$

5. Find the missing numbers.

a. $900 + \boxed{} = 914$	b. $620 + \boxed{} = 680$	c. $600 - \boxed{} = 570$
d. $\boxed{} - 20 = 40$	e. $\boxed{} - 70 = 70$	f. $572 - \boxed{} = 512$

6. Write the amounts using the dollar symbol and a decimal point.

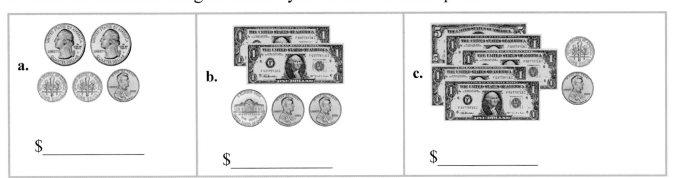

a.	b.	c.
$_____	$_____	$_____

7. Fay bought apples for $1.48 and gave the clerk $5.00.
   What was her change?

8. What is the total if you have six dimes,
   three nickels and a quarter?

9. Weigh yourself. I weigh _____. Now weigh yourself holding as many

   books as you can hold. I weigh _____ with the books.

   How much do the books weigh? _____

10. Solve.

a. Mia has saved $28. She wants to buy a sewing kit for $45. After she earns $13, can she buy it?	b. Find the cost of buying three rakes for $17 each.	c. One sack of potatoes weighs 22 kg. How much do four sacks weigh?
+	+	+

77

This page left blank intentionally.

Grade 2 End-of-the Year Test

Instructions

This test is quite long, because it contains lots of questions on all of the major topics covered in the *Math Mammoth Grade 2 Complete Curriculum*. Its main purpose is to be a diagnostic test—to find out what the student knows and does not know. The questions are quite basic and do not involve especially difficult word problems.

Since the test is so long, I do not recommend that you have the student do it in one sitting. You can break it into 3-5 parts and administer them on consecutive days, or perhaps on morning/evening/morning/evening. Use your judgment.

A calculator is not allowed.

My suggestion for grading is below. The total is 134 points. A score of 107 points is 80%.

Question	Max. points	Student score
Basic Addition and Subtraction Facts		
1	16 points	
2	16 points	
3	6 points	
	subtotal	/ 38
Mental Addition and Subtraction with Two-Digit Numbers and Word Problems		
4	1 point	
5	2 points	
6	3 points	
7	1 point	
8	3 points	
9	3 points	
10	6 points	
	subtotal	/ 19

Question	Max. points	Student score
Three-Digit Numbers		
11	2 points	
12	2 points	
13	2 points	
14	6 points	
15	4 points	
	subtotal	/ 16
Regrouping in Addition and Subtraction, Including Word Problems		
16	3 points	
17	4 points	
18	2 points	
19	2 points	
20	2 points	
21	3 points	
	subtotal	/ 16

Question	Max. points	Student score
Clock		
22	6 points	
23	5 points	
	subtotal	/ 11
Money		
24	2 points	
25	2 points	
26	2 points	
	subtotal	/ 6
Geometry and Measuring		
27	2 points	
28	4 points	
29	1 point	
30	4 points	
	subtotal	/ 11

Question	Max. points	Student score
Fractions		
31	4 points	
32	6 points	
	subtotal	/ 10
Concept of Multiplication		
33	2 points	
34	2 points	
35	3 points	
	subtotal	/ 7
	TOTAL	**/ 134**

End of Year Test - Grade 2

Basic Addition and Subtraction Facts

In problems 1 and 2, your teacher will read you the addition and subtraction questions. Try to answer them as quickly as possible. In each question, he/she will only wait a little while for you to answer, and if you don't say anything, your teacher will move on to the next problem. So just try your best to answer the questions as quickly as you can.

1. Add.

	a.	b.	c.	d.
	$6 + 7 =$ _____	$7 + 4 =$ _____	$8 + 8 =$ _____	$9 + 5 =$ _____
	$9 + 9 =$ _____	$5 + 8 =$ _____	$6 + 6 =$ _____	$7 + 7 =$ _____
	$5 + 6 =$ _____	$3 + 9 =$ _____	$2 + 9 =$ _____	$8 + 6 =$ _____
	$8 + 7 =$ _____	$5 + 7 =$ _____	$4 + 8 =$ _____	$8 + 9 =$ _____

2. Subtract.

	a.	b.	c.	d.
	$12 - 3 =$ _____	$11 - 3 =$ _____	$14 - 5 =$ _____	$13 - 4 =$ _____
	$15 - 7 =$ _____	$12 - 8 =$ _____	$12 - 4 =$ _____	$15 - 6 =$ _____
	$13 - 6 =$ _____	$14 - 6 =$ _____	$18 - 9 =$ _____	$12 - 6 =$ _____
	$11 - 7 =$ _____	$16 - 8 =$ _____	$16 - 7 =$ _____	$14 - 7 =$ _____

3. Fill in the missing numbers. The four problems form a fact family.

a. $2 + \boxed{} = 11$	b. ____ + ____ = 17	c. ____ + ____ = ____
$\boxed{} + 2 = 11$	____ + ____ = 17	____ + ____ = ____
$11 - 2 = \boxed{}$	$17 - 8 =$ ____	$12 -$ ____ $= 5$
$11 - \boxed{} = 2$	$17 -$ ____ $=$ ____	____ $-$ ____ $=$ ____

Mental Addition and Subtraction with Two-Digit Numbers and Word Problems

4. What is double 35?

5. Mary picked 5 apples and Bill picked 9. The children shared
   all of their apples evenly. How many did each child get?

6. List here the even numbers from 10 to 20.

7. Find the difference of 75 and 90.

8. Ed had saved $16. Then grandma gave him $10.
   Now how much more does he need in order to
   buy a toolset for $32?

9. Find the missing numbers.

   a. $82 + \underline{\hspace{1cm}} = 90$ **b.** $13 + \underline{\hspace{1cm}} = 21$ **c.** $90 - \underline{\hspace{1cm}} = 83$

10. Calculate mentally.

a. $59 + 8 = \underline{\hspace{1cm}}$	**b.** $52 + 40 = \underline{\hspace{1cm}}$	**c.** $76 - 50 = \underline{\hspace{1cm}}$
$62 + 8 = \underline{\hspace{1cm}}$	$45 + 9 = \underline{\hspace{1cm}}$	$54 - 23 = \underline{\hspace{1cm}}$

Three-Digit Numbers

11. Write with numbers.

   a. 6 tens 2 hundreds 7 ones = $\underline{\hspace{2cm}}$ **b.** 8 ones 9 hundreds = $\underline{\hspace{2cm}}$

12. Skip-count by tens.

   568, 578, $\underline{\hspace{1.5cm}}$, $\underline{\hspace{1.5cm}}$, $\underline{\hspace{1.5cm}}$, $\underline{\hspace{1.5cm}}$, $\underline{\hspace{1.5cm}}$

13. Write the numbers in order from the smallest to the greatest.

a. 417, 714, 447	b. 89, 998, 809

14. Calculate mentally.

a. $560 + 40 =$ _____ $560 + 400 =$ _____	b. $520 - 20 =$ _____ $520 - 200 =$ _____	c. $362 - 30 =$ _____ $362 - 300 =$ _____

15. Compare the expressions and write $<$, $>$, or $=$.

a. $100 - 5 - 3 \ \boxed{} \ 98 - 6$

b. $40 + 8 + 200 \ \boxed{} \ 20 + 800 + 4$

c. $50 + 120 \ \boxed{} \ 125$

d. $\frac{1}{2}$ of $800 \ \boxed{} \ 399 + 5$

Regrouping in Addition and Subtraction, including Word Problems

16. Add.

a.
$$\begin{array}{r} 3\ 5 \\ 3\ 6 \\ +\ 1\ 2 \\ \hline \end{array}$$

b.
$$\begin{array}{r} 2\ 2\ 4 \\ +\ 4\ 5\ 8 \\ \hline \end{array}$$

c.
$$\begin{array}{r} 4\ 3\ 8 \\ 1\ 7 \\ +\ 2\ 9\ 3 \\ \hline \end{array}$$

17. Subtract. Check by adding the result and what was subtracted.

a. $\begin{array}{r} 6\ 1 \\ -\ 3\ 7 \\ \hline \end{array}$ $+$ ____	b. $\begin{array}{r} 9\ 7\ 0 \\ -\ 2\ 4\ 8 \\ \hline \end{array}$ $+$ ____

18. Jennifer bought two vacuum cleaners for $152 each. What was the total cost?

19. A box contains 450 disks in all. Of them, 126 are music CDs and the rest are DVDs. How many DVDs are in the box?

20. The distance from Mark's home to his grandma's house is 218 miles. How many miles long is a round trip?

21. Every day Janet jogs around a rectangular-shaped jogging track. One side is 150 yards and another side is 300 yards.

    a. Mark the distances in the picture.

    b. Calculate what distance Janet goes when she jogs around it once.

Clock

22. Write the time with *hours:minutes*, and using "past" or "till".

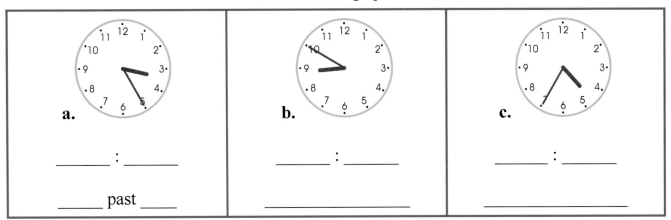

a.

_____ : _____

_____ past _____

b.

_____ : _____

_____

c.

_____ : _____

_____

23. How much time passes? Fill in the table.

from	3:00	2:00	1 AM	11 AM	8 PM
to	3:05	2:30	8 AM	1 PM	midnight
amount of time					

Money

24. How much money? Write the amount.

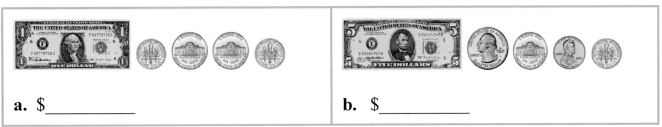

a. $_____

b. $_____

25. Find the change, if you buy a meal for $3.35
    and you pay with $4.

26. Bill bought an eraser that cost 85¢. He paid with $1.
    What was his change?

27. Identify the shapes.

    Shape A: _____

    Shape B: _____

28. **a.** Join the dots in order (A-B-C-D)
    with straight lines. Use a ruler.

    b. What shape is formed?

    _____

    c. Measure the sides of the shape to the nearest half-inch.

    Side AB: about _____   Side BC: about _____

    Side CD: about _____   Side DA: about _____

29. Measure this line to the nearest centimeter.

    ▬▬▬▬▬▬▬▬▬▬▬▬▬▬▬   about _____ cm

30. Which measuring unit or units could you use to find these amounts?
    Centimeter (cm), inch (in), meter (m), foot (ft), mile (mi), or kilometer (km)?
    Sometimes two different units are possible. If so, write both.

Distance	Unit(s)
how long my pencil is	
the distance from London to New York	
the height of a wall	
the distance it is to the neighbor's house	

Fractions

31. Divide these shapes. Then color as you are asked to.

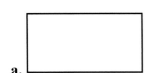

a.

Divide this into thirds. Color $\frac{2}{3}$.

b.

Divide this into halves. Color $\frac{1}{2}$.

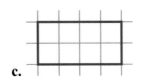

c.

Divide this into halves. Color $\frac{2}{2}$.

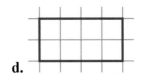

d.

Divide this into fourths. Color $\frac{3}{4}$.

32. Color. Then compare and write $<$, $>$, or $=$ between the fractions.

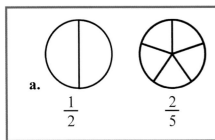

a.

$\frac{1}{2}$ $\frac{2}{5}$

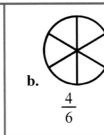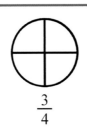

b.

$\frac{4}{6}$ $\frac{3}{4}$

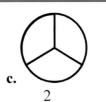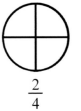

c.

$\frac{2}{3}$ $\frac{2}{4}$

Concept of Multiplication

33. Write a multiplication sentence for each picture.

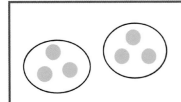

a. _____ × _____ = _____

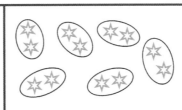

b. _____ × _____ = _____

34. Write a <u>multiplication</u> for each addition, and solve.

a. $5 + 5 + 5$	b. $4 + 4 + 4 + 4 + 4$
_____ × _____ = _____	_____ × _____ = _____

35. Solve.

a. $2 \times 5 =$ _____	b. $3 \times 3 =$ _____	c. $3 \times 10 =$ _____

Math Mammoth Grade 2 Review Workbook Answers

Some Old, Some New, p. 6

1.

a.	b.	c.	d.
$51 + 7 = 58$ $81 + 7 = 88$	$46 + 3 = 49$ $96 + 3 = 99$	$72 + 5 = 77$ $32 + 5 = 37$	$35 + 5 = 40$ $95 + 5 = 100$

2.

a.	b.	c.	d.
$49 - 5 = 44$ $89 - 5 = 84$	$29 - 3 = 26$ $69 - 3 = 66$	$60 - 7 = 53$ $80 - 7 = 73$	$38 - 4 = 34$ $78 - 4 = 74$

3. a. $20 + $20 + $20 = $60   b. His bill is $39. $32 + $1 + $2 + 4 = $39.

4.

a.	b.	c.	d.
$21 + 40 = 61$ $56 + 30 = 86$	$40 + 23 = 63$ $20 + 78 = 98$	$72 - 50 = 22$ $66 - 40 = 26$	$89 - 30 = 59$ $45 - 20 = 25$

5.

W	E	L	L	D	O	N	E
1st	5th	9th	9th	4th	2nd	3rd	5th

6.

a.	b.	c.
$2 + 8 = 10$ $8 + 2 = 10$ $10 - 8 = 2$ $10 - 2 = 8$	$7 + 2 = 9$ $2 + 7 = 9$ $9 - 7 = 2$ $9 - 2 = 7$	$5 + 3 = 8$ $3 + 5 = 8$ $8 - 3 = 5$ $8 - 5 = 3$

7. a. $16 - 8 = 8$
   b. $9 - 5 = 4$
   c. $60 - 30 = 30$

8. The even numbers are 72, 60, and 8.

9.

a. $\dfrac{1}{2}$ of 50 is ___25___.

b. $\dfrac{1}{2}$ of 88 is ___44___.

c. $\dfrac{1}{2}$ of 46 is ___23___.

10. Each boy got nine cars.

11. She had 26 potatoes.

12. They have 39 colored pencils together. (Tina has 26.)

Some Old, Some New - Test, p. 8

1. a. 60, 52
   b. 37, 90
   c. 56, 35
   d. 85, 28

2. a.  b.

3.

a. 2 + 8 = 10	b. 6 + 3 = 9	c. 7 + 8 = 15
8 + 2 = 10	3 + 6 = 9	8 + 7 = 15
10 − 8 = 2	9 − 6 = 3	15 − 7 = 8
10 − 2 = 8	9 − 3 = 6	15 − 8 = 7

4. She read 46 pages.

5.

Number	Even?	Odd?
4	X	
10	X	

Number	Even?	Odd?
9		X
16	X	

Number	Even?	Odd?
11		X
18	X	

Clock Review, p. 9

1.

a. 1:50 / 10 till 2	b. 4:25 / 25 past 4	c. 8:55 / 5 till 9	d. 11:05 / 5 past 11
e. 3:40 / 20 till 4	f. 7:25 / 25 past 7	g. 5:30 / half past 5	h. 12:00 / 12 o'clock

2.

Time now	2:30	6:55
5 min. later	2:35	7:00

Time now	9:05	5:40
10 min. later	9:15	5:50

3. He works eight hours.

4. The class ends at 12 o'clock.

5. He goes to the chess club next time on March 24.

6. November

Clock Test, p. 10

1.

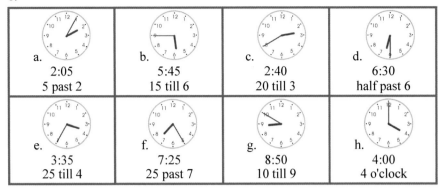

| a. 2:05 5 past 2 | b. 5:45 15 till 6 | c. 2:40 20 till 3 | d. 6:30 half past 6 |
| e. 3:35 25 till 4 | f. 7:25 25 past 7 | g. 8:50 10 till 9 | h. 4:00 4 o'clock |

2.

Time now	3:50	7:25
5 minutes later	3:55	7:30

Time now	9 AM	12 noon
1 hour later	10 AM	1 PM

3.

a. 20 past 4 4:20	b. 15 past 11 11:15	c. 15 till 12 11:45	d. 25 till 7 6:35

4.

from	5 AM	8 AM	2 AM	10 AM	11 AM
to	12 noon	2 PM	3 PM	10 PM	6 PM
hours	7	6	13	12	7

Mixed Review 1 , p. 11

1. a. 8 hours   b. 24   c. $9.00 now   d. $40.00

2.

a.
```
    40        30
  ↑+30 ↓+10 ↑−20 ↓−20
10        50        10
```
b.
```
    20        10
  ↑−40 ↓+30 ↑−40 ↓+20
60        50        30
```

```
10 + 10 = 20
15 + 15 = 30
20 + 20 = 40
25 + 25 = 50
30 + 30 = 60
35 + 35 = 70
40 + 40 = 80
```

3.

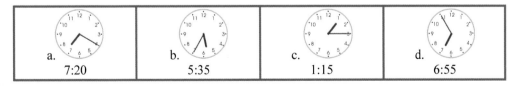

| a. 7:20 | b. 5:35 | c. 1:15 | d. 6:55 |

4. a. 3:40 and 4:15   b. Jim walked his dog for one hour.   c. 5:30 AM

5.

a. 3 + 6 = 9 6 + 3 = 9 9 − 3 = 6 9 − 6 = 3	b. 4 + 6 = 10 6 + 4 = 10 10 − 4 = 6 10 − 6 = 4	c. 3 + 5 = 8 5 + 3 = 8 8 − 5 = 3 8 − 3 = 5

6. A - HORSE

1.

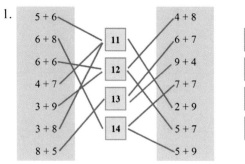 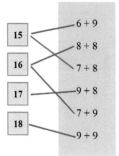

2.

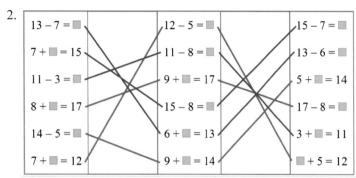

3. a. 7   b. 43   c. 7   d. 9

4. a. 7   b. 7   c. 7   d. 8   e. 6   f. 6   g. 5   h. 5   i. 8

5. 75, 70, 65, 63, 60, 54, 51

6. a.

Cookies you have	Cookies your friend has	Together we have	even/odd	Can you share evenly?
3	5	8	even	yes
5	9	14	even	yes
9	3	12	even	yes
9	7	16	even	yes

b.

Cookies you have	Cookies your friend has	Together we have	even/odd	Can you share evenly?
5	6	11	odd	no
7	8	15	odd	no
9	4	13	odd	no
1	12	13	odd	no

7. IT GOT HOT IN THE HEAT.

8. a. Jane has seven more than Jack. 20 − 13 = 7 or 13 + ___ = 20
   b. Sofia has 11. 14 − 3 = 11.
   c. Jacob has 7 pawns. 5 + 2 = 7.
   d. You will need to save $8 more. Think: $20 + ___ = $28.
      After the neighbor pays you, you still need $6. You have $20 + $2 = $22. Think: $22 + _6_ = $28.
   e. I need seven more squares to get to the end of the game. You roll 5 + 6 = 11, and 11 + _7_ = 18.
      To get to the end, you need to roll a total of seven on two dice. You could roll 3 and 4, or 1 and 6, or 2 and 5.

Addition and Subtraction Facts Within 0-18 Test, p. 16

1. a. 15, 13   b. 17, 12   c. 7, 9
   d. 3, 9   e. 13, 13   f. 14, 15

2.

a. 14 − 5 = <u>9</u>	b. 11 − 8 = <u>3</u>	c. 17 − 8 = <u>9</u>
<u>9</u> + <u>5</u> = 14	<u>3</u> + <u>8</u> = 11	<u>9</u> + <u>8</u> = <u>17</u>

3. a. =   b. >   c. >

4. a. 5, 8   b. 8, 6   c. 7, 7

5. a. 16 teddy bears. 7 + 9 = 16
   b. I still need to save $2. $7 + $5 = $12. And $12 + $2 = $14.

Mixed Review 2, p. 17

1. a. 2:50   b. 7:25   c. 8:55   d. 11:50

2.

	a. 2 : 35	b. 12 : 40	c. 7 : 30	d. 3 : 55
5 min. later →	2 : 40	12 : 45	7 : 35	4 : 00

3.

from	2:25	2:20	7:00	11:30	6:05
to	2:35	2:40	7:15	11:50	6:15
minutes	<u>10 minutes</u>	20 minutes	15 minutes	20 minutes	10 minutes

4. Each child got 9 raisins and 6 almonds.

5. a. 10 = 5 + 5   b. 16 = 8 + 8   c. 40 = 20 + 20

6.

7. a. Isabella ate 10 strawberries. Together, they ate 30 strawberries.
   b. Kyle had $20.
   c. Jane ate 22 strawberries.
   d. Emily is 26 years older than Hannah.
   e. Ann has more cars. She has four more cars.
   f. Jim has $13.

Mixed Review 3, p. 19

1. a. 13, 17 b. 16, 13 c. 9, 8 d. 7, 7

2.

from	9 AM	6 AM	11 AM	12 AM	10 AM
to	1 PM	8 PM	4 PM	12 PM	2 PM
hours	4	14	5	12	4

3. a. 5 Tuesdays b. Jan 20

4. a. She started at 12 noon.
   b. Grandma sleeps six hours.

5. a. 97, 34 b. 68, 23 c. 94, 38

6.

a. $18 + 4 = 22$	b. $75 + 5 = 80$	c. $56 + 3 = 59$
d. The difference of 8 and 12 is 4.	e. The difference of 43 and 49 is 6.	f. The difference of 21 and 30 is 9.

7. a. 5, 4 b. 5, 10 c. 3, 3

8.

a. $8 + 6 = 14$ $14 - 8 = 6$	b. $5 + 9 = 14$ $14 - 5 = 9$	c. $6 + 6 = 12$ $12 - 6 = 6$

9. a. 5, 8, 3 b. 6, 4, 6 c. 5, 5, 7 d. 8, 5, 6

10. $7 + 6 = 13$ $13 - 6 = 7$
    $6 + 7 = 13$ $13 - 7 = 6$

Regrouping in Addition Review, p. 21

1. a. 27, 52 b. 36, 87 c. 64, 75

2. a. 35, 61 b. 27, 81 c. 92, 95

3.

a. $36 + 22$ $30 + 20 + 6 + 2$ $50 + 8 = 58$	b. $72 + 18$ $70 + 10 + 2 + 8$ $80 + 10 = 90$
c. $54 + 37$ $50 + 30 + 4 + 7$ $80 + 11 = 91$	d. $24 + 55$ $20 + 50 + 4 + 5$ $70 + 9 = 79$

4. a. Ted earned $50. $25 + $25 = $50. Together they earned $75. $50 + $25 = $75.
   b. Leah has 12. Half of 24 is 12.

5. a. 71 b. 72 c. 71 d. 61 e. 72
   f. 93 g. 95 h. 121 i. 117 j. 117

6. a. The total cost was $73. $18 + $15 + 40 = $73.
   b. The total bill is $102. $34 + $34 + $34 = $102.
   c. There are 106 stickers in total. 29 + 29 + 22 + 26 = 106.
   d. No, he cannot. $47 + $15 = $62, which is less than $65. He needs $3 more.

Regrouping in Addition Test, p. 23

1. a. 85   b. 122   c. 87   d. 61   e. 104

2. a. 59, 23   b. 73, 35   c. 93, 51

3. a. 23, 24   b. 59, 76

4. a. She worked 15 more hours. $28 - 13 = 15$
   b. The total cost is $46. $12 + 17 + 17 = 46$
   c. There are now 22 birds in the tree. $15 + 9 - 2 = 22$

Mixed Review 4, p. 24

1. a. 3   b. 5   c. 4   d. 12   e. 20   f. 16

2. a. 7   b. 20   c. 40   d. 8   e. 25   f. 91

3.
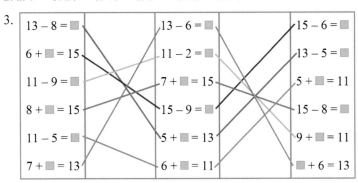

4. a. 6, 7   b. 4, 7   c. 6, 8   d. 8, 8

5. a. 3:55   b. 10:05   c. 3:45   d. 12:40

6. a. 4:05   b. 10:15   c. 3:55   d. 12:50

7. a. Shaun got the most points. Shaun got $14 + 14 = 28$ points.
   b. My brother is 8 years older than I am. $8 + 8 = 16$; $16 - 8 = 8$.
   c. Emma got 38 points. $31 + 7 = 38$.
   d. It costs $32. $26 + 6 = \$32$.
   e. Eight birds. $15 - 7 = 8$.
   f. Each girl gets 9 markers. $12 + 6 = 18$. Half of 18 is 9.

Mixed Review 5, p. 26

1. a. There are fourteen boys in the class.
   b. Andy has $16. $8 + 8 = 16$
      Together they have $24. $16 + 8 = 24$

2. a. 32, 68   b. 48, 59   c. 53, 28

3. The camera costs $134 now. $67 + 67 = \$134$

4. a. 25 past 7   b. 10 past 5   c. 10 till 6   d. 20 till 1   e. half past 12   f. 11 o'clock

5. a. 81   b. 35   c. 81   d. 120

6. a. It took Chris 60 minutes to make the cards.
   b. He finished at 1 o'clock.

7. a. 32   b. 9   c. 8   d. 11   e. 27   f. 36

8. The total cost is $18 + $18 + $18 + $25 = $79.

1. A hexagon

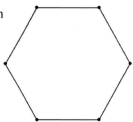

2. Answers vary. For example:

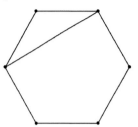

3.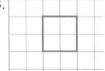

4.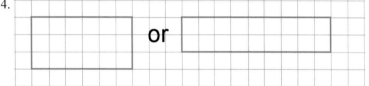

5. A cube. It has 6 faces. The faces are in the shape of a square.

6. She got a quadrilateral (to be exact, a parallelogram).

7. a. box   b. pyramid   c. cone

8.

 a. $1 = \frac{4}{4}$ b. $1 = \frac{3}{3}$

9.

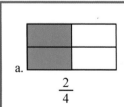

a. $\frac{2}{4}$ b. $\frac{1}{3}$ c. $\frac{2}{3}$ d. $\frac{2}{2}$

10.

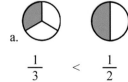

a. $\frac{1}{3} < \frac{1}{2}$ b. $\frac{2}{3} < \frac{3}{4}$ c. 1 whole $> \frac{3}{4}$

Geometry and Fractions Test, p. 30

1.

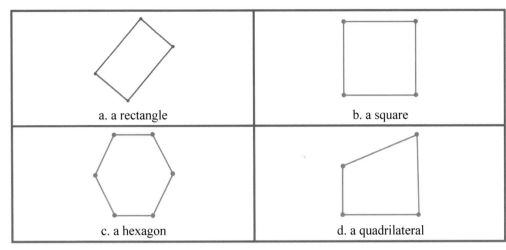

2.

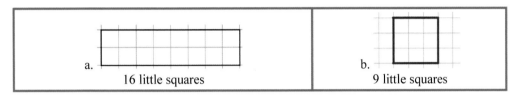

3. Answers will vary. Please check the student's work.

4. a. 1/5   b. 5/6   c. 3/3   d. 2/4

5.

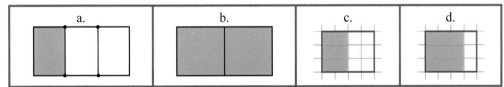

6.

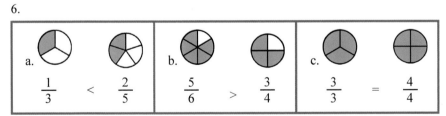

Mixed Review 6, p. 32

1. a. 5   b. 20   c. 8   d. 52

2. a. 3   b. 5   c. 2

3. a. 9   b. 5   c. 7   d. 12   e. 11   f. 13

4. a. 11, 21   b. 12, 92   c. 15, 55   d. 53   e. 56   f. 93

5. a. 104   b. 97   c. 95   d. 107   e. 96

6. a. $55   b. $112   c. $116

Mixed Review 6, cont.

7.

a. $\boxed{8} + \boxed{8} + 2 + 8$ $= \boxed{16} + 10$ $= 26$	b. $9 + 5 + 5 + 8$ $= 17 + 10$ $= 27$	c. $6 + 7 + 3 + 5$ $= 11 + 10$ $= 21$
d. $7 + 7 + 8 + 8$ $= 30$	e. $9 + 4 + 4 + 7$ $= 24$	f. $6 + 4 + 4 + 9$ $= 23$

8. a. It costs \$101. \$78 + \$23 = \$101
   b. The first shirt costs \$14 more. \$29 − \$15 = \$14.
   c. They cost together \$44. \$29 + \$15 = \$44.

Mixed Review 7, p. 34

1.

a. $\boxed{T}$ $\boxed{T}$ and $\boxed{T}\boxed{T}$ $\boxed{T}\boxed{T}$ $20 + 40 = 60$ $40 + 20 = 60$ $60 - 40 = 20$ $60 - 20 = 40$	b. $\boxed{T}\boxed{T}$ $\bullet\bullet$ $\bullet\bullet$ and $\bullet\bullet\bullet$ $24 + 7 = 31$ $7 + 24 = 31$ $31 - 7 = 24$ $31 - 24 = 7$

2. a. 119   b. 139   c. 107   d. 78

3. a. LLAMA   b. ORDINAL

4.

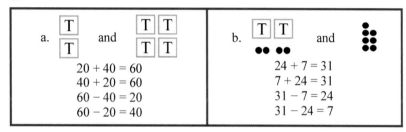

	a. 2:55	b. 7:10	c. 2:25	d. 11:00
5 min. later →	3:00	7:15	2 :30	11:05

5. There are six blue flowers. $15 - 4 - 5 = 6$

6. a. $16 < 17$ b. $22 > 21$ c. $8 < 16$ d. $42 < 43$ e. $34 > 31$ f. $7 < 8$

7.

a. $36 + 22$ $\boxed{30 + 20} + \boxed{6 + 2}$ $\quad 50 \quad + \quad 8 \quad = 58$	b. $72 + 18$ $\boxed{70 + 10} + \boxed{2 + 8}$ $\quad 80 \quad + \quad 10 \quad = 90$
c. $54 + 37$ $80 + 11 = 91$	d. $24 + 55$ $70 + 9 = 79$

8. a. | 464 | 474 | 484 | 494 | 504 | 514 | 524 | 534 | 544 |

   b. | 400 | 450 | 500 | 550 | 600 | 650 | 700 | 750 | 800 |

Three-Digit Numbers Review p. 36

1. a. 486   b. 487   c. 496   d. 586

2. a. 178, 179, 180   b. 200, 201, 202   c. 799, 800, 801   d. 916, 917, 918

3. a. 709   b. 674   c. 580   d. 558

4. 695, 700, 705, 710, 715, 720

5. a. 282, 292, 302   b. 535, 545, 555

6.

200	220	240	260	280
300	320	340	360	380
400	420	440	460	480

7.

a. 238 < 265	b. 391 > 193	c. 405 < 450	d. 981 > 819
e. 8 + 600 < 60 + 800		f. 30 + 300 + 5 > 90 + 8 + 100	

8.

a. 109, 199, 901	b. 175, 177, 717

9. a. 323 + 40 = 363   b. 262 + 300 = 562

10. a. 920, 480   b. 908, 270   c. 719, 54

11. a. 600, 690   b. 929, 453   c. 542, 814

12.

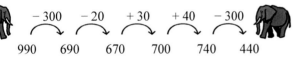

│ − 300 │ − 20 │ + 30 │ + 40 │ − 300 │

990   690   670   700   740   440

13. a. 260 sheep. 250 − 10 + 20 = 260   b. 140 goldfish. 170 − 30 = 140
   c. Now he has 190 goldfish. 140 + 50 = 190. He has 100 rainbow fish. 30 + 70 = 100
   d. She traveled 860 km. 400 + 30 + 400 + 30 = 860

Puzzle corner: a. 32   b. 8

Three-Digit Numbers Test, p. 39

1. a. 475, 480, 485, 490, 495, 500, 505   b. 376, 386, 396, 406, 416, 426, 436

2. a. 235 = 200 + 30 + 5   b. 805 = 800 + 0 + 5

3. a. 688, 460   b. 285, 106

4. a. <   b. >   c. <   d. >

5. a. 689 < 869 < 986   b. 245 < 452 < 524

6. a. 256 = 256   b. 809 < 890   c. 462 > 246   d. 703 < 706

7. a. △ = 40   b. △ = 8   c. △ = 500

8. a. 565, 248   b. 402, 380   c. 278, 94

9. a. Natalie counted 25 cars b. Jayden counted 35 cars. c. Natalie counted five more cars than Caleb.

1.

	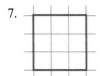 a. 2 : 15	b. 8 : 50	c. 11 : 05	d. 9 : 55
5 min. later →	2 : 20	8 : 55	11 : 10	10 : 00

2. a. 7   b. 8   c. 3
   d. 14   e. 12   f. 18

3. a. 13, 23   b. 17, 77   c. 13, 43

4. John has five plums left. John got 8 (half of 16). Then, 8 − 3 = 5.
   Jane has six plums left. Jane got 8 (half of 16). Then, 8 − 2 = 6.

5. a. 6   b. 9   c. 50
   d. 20   e. 7   f. 300

6. a. 125   b. 104   c. 104   d. 125   e. 104

7.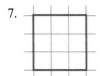

8.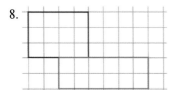

9.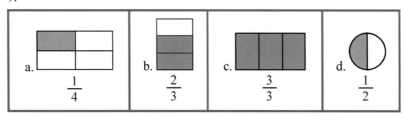

10.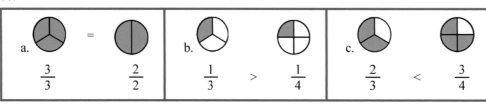

Mixed Review 9, p. 43

1. a. first b. fourth c. fourth d. second

2. He spends two hours practicing.

3. The total cost is $29.   12 + 12 + 5 = 29

4. a. 98   b. 145   c. 143   d. 88   e. 82

5.

a. 66 + 4 = 70 92 + 8 = 100	b. 31 + 3 + 6 = 40 63 + 2 + 5 = 70	c. 47 + 2 + 1 = 50 32 + 6 + 2 = 40

6.

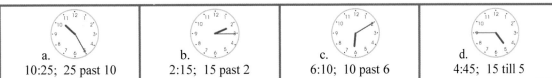

a. 10:25; 25 past 10	b. 2:15; 15 past 2	c. 6:10; 10 past 6	d. 4:45; 15 till 5

7. a hexagon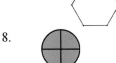

8.

   a. $1 = \dfrac{4}{4}$ b. $1 = \dfrac{6}{6}$ c. $1 = \dfrac{5}{5}$

9.

a.	b.	c.	d.
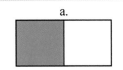			

Measuring Review, p. 45

1.

Distance	Unit or units
from your house to the grocery store	mi or km
from Miami to New York	mi or km
the distance across the room	m or ft
the height of a bookcase	ft, in, m, or cm

2. about 6 cm _or_ about 2 1/2 in.

3. a. ▬▬▬▬▬▬▬▬▬▬▬▬▬▬▬▬

   b. ▬▬▬▬▬▬▬▬▬▬▬▬▬▬▬

4. The longer pencil is about 7 cm longer than the shorter one.
   The longer pencil is about 2.5 inches longer than the shorter one.

5. Answers will vary.

Measuring Test, p. 46

1. (a), (c) and (d) do not make sense.

2. #1 is about 4 inches, 10 cm   #2 is about 5 inches, 13 cm

3. a. ▬▬▬▬▬▬▬▬▬▬▬▬▬▬▬▬▬▬

   b. ▬▬▬▬▬▬▬▬▬▬▬▬▬▬▬

4. centimeter, inch, foot, kilometer

5. 3 meters

6.

Distance	Unit
from Florida to California	km
around your head	cm

Distance	Unit
length of a garden	m
height of a room	m

7. Answers will vary.

Mixed Review 10, p. 47

1.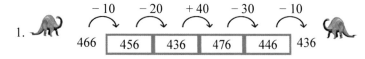

466 | 456 | 436 | 476 | 446 | 436

-10 -20 $+40$ -30 -10

2. a. △ = 30   b. △ = 9   c. △ = 200

3. a. 806, 816, 826, 836, 846, 856, 866, 876   b. 542, 532, 522, 512, 502, 492, 482, 472

4. a. 280, 285, 290, 295, 300, 305, 310, 315   b. 1000, 995, 990, 985, 980, 975, 970, 965

5. a. 300   b. 494   c. 998

6. a. ▬▬▬▬▬▬▬▬▬▬▬▬▬▬

   b. exceeds the width of the paper

   c. ▬▬▬▬▬▬▬▬▬▬▬▬▬▬▬▬▬

   d. ▬▬▬▬▬▬▬▬▬

7. 90 + 90 = 180 km

8.

Distance	Unit
from New York to Chicago	km
the length of your room	m
the length of a pencil	cm

Distance	Unit
around your neck	cm
the width of a butterfly	cm
how far you can throw a ball	m

9. a. 29   b. 480   c. 740

10.

a. 10 past 8 8 : 10	b. 15 till 7 6 : 45	c. 25 past 12 12 : 25	d. half-past 7 7 : 30
e. 9 o'clock 9 : 00	f. 20 till 6 5 : 40	g. 5 till 11 10 : 55	h. 25 till 4 3 : 35

11. Answers will vary.

12. a. triangles   b. hexagons

Puzzle corner: 119, 128, 137, 146, 155, 164, 173, 182, 191

Mixed Review 11, p. 50

1. a. She needs to buy six more apples. $8 + 8 - 10 = 6$
   b. Two go-karts would cost $36. $18 + 18 = 36$
   c. She has eight cats and there are four kittens. $4 + 4 = 8$

2.

     a. 1:25   b. 3:15   c. 15 till 3

3.

a. $577 - 10 = 567$	b. $926 - 0 = 926$
$577 - 20 = 557$	$926 - 100 = 826$
$577 - 30 = 547$	$926 - 200 = 726$
$577 - 40 = 537$	$926 - 300 = 626$
$577 - 50 = 527$	$926 - 400 = 526$
$577 - 60 = 517$	$926 - 500 = 426$

4. a. 17, 11, 14   b. 16, 12, 16   c. 13, 11, 14

5. a. a pentagon   b. a hexagon   c. a triangle

6. There are 18 squares in the grid.

7.

Ordinal Number	Name	Ordinal Number	Name
1st	first	8th	eighth
2nd	second	9th	ninth
3rd	third	10th	tenth
4th	fourth	11th	eleventh
5th	fifth	12th	twelfth
6th	sixth	13th	thirteenth
7th	seventh	14th	fourteenth

Regrouping in Addition and Subtraction Review, p. 52

1. a. 692   b. 417   c. 718   d. 764

2. The three bicycles cost a total of $462. Add: $154 + 154 + 154 = 462$.

3. a. 120, 820   b. 180, 290   c. 740, 550

4. It is 380 feet all of the way around. Add $120 + 70 + 120 + 70 = 380$.

5. a. 34, 88   b. 15, 63   c. 35, 84   d. 723, 882   e. 165, 556   f. 304, 550

6. a. 8, 48   b. 8, 88   c. 5, 4   d. 9, 5   e. 16, 13   f. 43, 29

Regrouping in Addition and Subtraction Review, cont.

7. a.
```
   2 5 4
 + 4 7 7
 -------
   7 3 1
```
b.
```
   5 8 9
 + 3 2 5
 -------
   9 1 4
```
c.
```
   2 0 6
 + 6 8 6
 -------
   8 9 2
```
d.
```
   6 8 1
 + 2 1 9
 -------
   9 0 0
```

8. a. 39 people. 52 − 13 = 39
   b. 39 toys. 23 + 16 = 39
   c. 17 toys. 33 − 16 = 17
   d. 700 points. 465 + 145 + 90 = 700
   e. 40 jumping jacks. 26 + 14 = 40.

9. a.

CHILD	POINTS
Charlie	15
Bill	24
Amy	28
Cindy	21
Sarah	19

   b. 4 points
   c. 6 points

Puzzle corner:

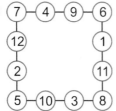

Regrouping in Addition and Subtraction Test, p. 56

1. a. 654   b. 937   c. 782   d. 35

2. a. 737, 964   b. 376, 748

3. a. 120, 330
   b. 750, 54
   c. 41, 2

4. Nancy's total bill was $387.

5. a. There are 182 sacks of wheat left. 250 − 68 = 182
   b. There are 19 black kittens. 52 − 15 − 18 = 19
   c. The three bags weigh 80 pounds. 15 + 15 + 50 = 80
   d. They sold 28 coffee makers in February. 47 − 19 = 28
      They sold 75 coffee makers in two months. 47 + 28 = 75
   e. Grandpa walked a total of 720 meters. 300 + 300 + 120 = 720

Mixed Review 12, p. 58

1.

8	9	10	11	12	13	14	15	16	17
16	18	20	22	24	26	28	30	32	34

2. a. Each girl got 15 marbles.
   b. There are 14 kilograms of potatoes left.
   c. Katy now has $21. Half of $60 is $30. Then, $30 − $9 = $21.
   d. Mom had 16 apples to start with.

Mixed Review 12, cont.

3. a. 139   b. 471   c. 204   d. 336   e. 761   f. 950

4. a. These months have 31 days: January, March, May, July, August, October, and December.
   b. These months have 30 days: April, June, September, and November.
   c. February has 28 days, except every leap year, and then it has 29 days.

5.

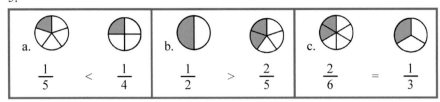

6. a. △ = 80   b. △ = 90   c. △ = 160
   a. Subtract 800 and 720. That works because subtraction is the opposite operation of addition.
   b. Subtract 200 and 110. The reason being, when the subtrahend is missing, it is like one "part" is missing.
      To find that part, subtract the other part from the total.
   c. Add 70 and 90. The minuend is missing, which is the "total" in the subtraction, so to find it, add the "parts."

7.

a.	b.	c.
25 past 3 3 : 25	5 past 12 12 : 05	10 till 9 8 : 50

8. Check the student's work.

Mixed Review 13, p. 60

1. a. 11   b. 22   c. 37   d. 37

2. a. 9, 16
   b. 12, 15
   c. 14, 15
   d. 18, 13

3.

a. $17 - 11 = 6$
   Think: $11 + 6 = 17$
	b. $43 - 37 = 6$
	   Think: $37 + 6 = 43$
		c. $66 - 59 = 7$
		   Think: $59 + 7 = 66$

d. $35 - 28 = 7$	e. $80 - 77 = 3$	f. $100 - 94 = 6$

4.

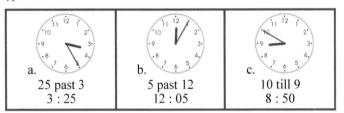

| -4 | -4 | -3 | -3 | -6 | -2 | -5 |

79   75   71   68   65   59   57   52

5.

a. 15 past 6 6:15	b. 20 till 3 2:40	c. 5 past 10 10:05	d. half past 3 3:30
e. 15 till 8 7:45	f. 20 till 12 11:40	g. 5 till 1 12:55	h. 25 past 1 1:25

Mixed Review 13, cont.

6. a. Dan's sister weighs 118 pounds. $138 - 20 = 118$
   b. Together they weigh 256 pounds. $138 + 118 = 256$

7. a - b.
   c. triangles

8. a. $\frac{2}{6}$ $\frac{2}{3}$ $\frac{2}{4}$ $\frac{2}{5}$

   b. 2/3 is the most pie to eat.

9. a. <   b. >   c. $789 < 798$

Money Review, p. 62

1. a. 33¢   b. 50¢

2. a. One quarter and three pennies.
   b. Three quarters, one dime, one nickel, and three pennies.

3. a. $2.35   b. $7.19   c. $0.45   d. $0.49

4. a. You give 70¢, and your change is 5¢.
   b. You give $1.00, and your change is 8¢

5.
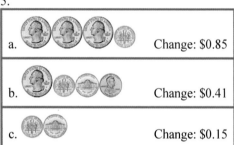

a.	Change: $0.85
b.	Change: $0.41
c.	Change: $0.15

6. Alex has 77¢. Together they have $2.03.

7. a. $1.76   b. $3.61

Money Test, p. 64

1. a. $1.20   b. $5.36   c. $5.71   d. $2.30

2. a. $0.12   b. $2.15

3. a. $1.56 + 1.56 + 0.78 = 3.90
   b. $2.55 + 2.55 = 5.10

Mixed Review 14, p. 65

1. a. 316   b. 433   c. 643

2. a. Together they have 143 marbles.
   b. Mom has $80 left.
   c. He has 72 pages left to read.

Mixed Review 14, cont.

3. a. 65, 93   b. 283

4. The chart below lists the base side of the triangle first.

Triangle	in inches	in centimeters
Side 1	2 in	5 cm
Side 2	3 in	8 cm
Side 3	3 in	8 cm

5. a. Chris   b. 5 more   c. 4 fewer fish   d. 15 together

6. a. 17 cars
   b. 8 cars
   c. 35 cars
   d. Yes, he will have 26 and Tony 25.

Puzzle corner:

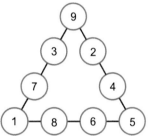

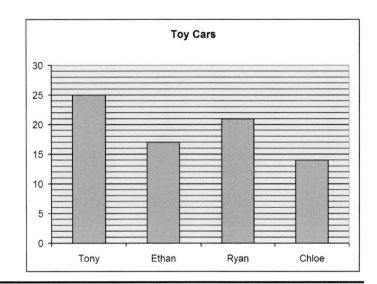

Mixed Review 15, p. 68

1.

   a. $\dfrac{1}{2}$ b. $\dfrac{1}{4}$ c. $\dfrac{3}{4}$ d. $\dfrac{2}{4}$ e. $\dfrac{4}{4}$ f. $\dfrac{2}{2}$

2.

a. From 3:00 to 8:00 <u>5</u> hours	d. From noon till 4 PM <u>4</u> hours
b. From 6 AM to 1 PM <u>7</u> hours	e. From 7 PM to 11 PM <u>4</u> hours
c. From 7 PM to midnight <u>5</u> hours	f. From 10 AM to 2 PM <u>4</u> hours

3. a. 15, 13, 6   b. 5, 6, 7   c. 7, 5, 7

Mixed Review 15, cont.

4. a. $0.94   b. $3.14   c. $8.22

5. a. It as five corners.   b. a pentagon

6. He has 10 marbles. $15 - 5 = 10$
   Together they have 25 marbles. $15 + 10 = 25$

7. Eva ate six pieces of pie. $4 + 2 = 6$

8. a. 537, 807   b. 600, 240   c. 532, 352

9. a. 332, 670   b. 270, 541

10.

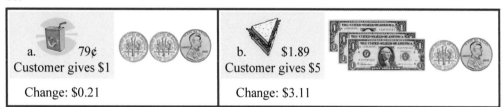

Exploring Multiplication Review, p. 70

1. a. $5 \times 1 = 5$ ▣▣▣▣▣

   b. $2 \times 10 = 20$

   c. $3 \times 2 = 6$

2. a. 18

   b. 12

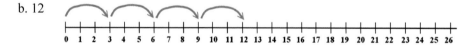

3. a. $3 + 3 + 3 = 9$ b. $2 + 2 + 2 + 2 = 8$

4. a. 6, 20, 40   b. 9, 60, 7   c. 3, 0, 20

5. a. 12   b. 8   c. 8   d. 4   e. 12   f. 4

6. $5 + 5 + 5 + 5 = 20$; $5 + 5 + 5 = 15$; $5 + 5 = 10$; $1 \times 5 = 5$; $2 \times 5 = 10$; $3 \times 5 = 15$; $4 \times 5 = 20$;
   $20 - 5 = 15$; $20 - 5 - 5 = 10$; $20 - 5 - 5 - 5 = 5$; $20 - 5 - 5 - 5 - 5 = 0$;

7. a. Fifty flowers. $5 \times 10 = 50$
   b. Jim has more. Now he has ten more.

8.

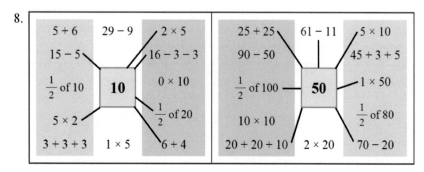

108

Exploring Multiplication Test, p. 72

1. a.
$6 \times 1 = 6$

   b.
$2 \times 7 = 14$

   c.
$3 \times 3 = 9$

2. a. 4×6
   b. 3×50

3. a. $8 + 8$
   b. $3 + 3 + 3 + 3 + 3$

4. a.

   b.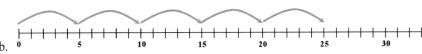

5. a. 12, 30   b. 0, 40   c. 6, 18
   d. 12, 16   e. 24, 15   f. 40, 800

Mixed Review 16, p. 73

1. a. 624, 244   b. 707, 277   c. 72, 4

2. a. $1.45   b. $5.21

3. Brett has 100¢. Together they have 157¢.

4. a. The total cost is $1.10 and the change is $3.90.
   b. The total cost is $1.60 and the change is $0.40.

5.

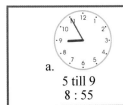		
a.	b.	c.
5 till 9	25 past 11	20 till 3
8 : 55	11 : 25	2 : 40

6. a. 9:00   b. 11:30   c. 2:45

7. a. 425, 652   b. 363, 548

8. 350 meters all the way around.

9. a. △ = 40   b. △ = 90   c. △ = 120

10. a. 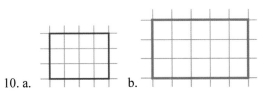   b.

11. a. 6 faces, squares
    b. 6 faces, rectangles

12. a. cone   b. pyramid
    c. cylinder   d. cube

Mixed Review 17, p. 76

1.

1	2	3	4	5	6	7	8	9	10
11	12	13	14	15	16	17	18	19	20
21	22	23	24	25	26	27	28	29	30

2. a. I will return January 28.
   b. 2, 9, 16, 23, 30

3. She now has $20.

4. a. 729   b. 440   c. 191   d. 524

5. a. 14   b. 60   c. 30   d. 60   e. 140   f. 60

6. a. $0.71   b. $2.07   c. $9.11

7. Her change was $3.52. $5.00 - 1.48 = 3.52$

8. The total is 100 cents or one dollar.

9. Answers will vary.

10. a. No, she still needs four more dollars. $45 - 28 - 13 = 4$
    b. The total cost of three rakes is $51. $17 + 17 + 17 = 51$
    c. Four sacks of potatoes would weigh 88 kilograms. $22 + 22 + 22 + 22 = 88$

End of the Year Test, p. 80

1. a. 13, 18, 11, 15
   b. 11, 13, 12, 12
   c. 16, 12, 11, 12
   d. 14, 14, 14, 17

2. a. 9, 8, 7, 4
   b. 8, 4, 8, 8
   c. 9, 8, 9, 9
   d. 9, 9, 6, 7

3.

a. $2 + 9 = 11$	b. $8 + 9 = 17$	c. $5 + 7 = 12$
$9 + 2 = 11$	$9 + 8 = 17$	$7 + 5 = 12$
$11 - 2 = 9$	$17 - 8 = 9$	$12 - 7 = 5$
$11 - 9 = 2$	$17 - 9 = 8$	$12 - 5 = 7$

4. $35 + 35 = 70$

5. $5 + 9 = 14$ They each got seven apples.

6. 10, 12, 14, 16, 18, 20

7. $90 - 75 = 15$

8. $32 - 16 - 10 = 6$ He needs six more dollars.

9. a. 8   b. 8   c. 7

10 a. 67, 70
   b. 92, 54
   c. 26, 31

11. a. 267   b. 908

12. 568, 578, 588, 598, 608, 618, 628

13. a. 417, 447, 714
    b. 89, 809, 998

14. a. 600, 960
    b. 500, 320
    c. 332, 62

15. a. 92 = 92   b. 248 < 824
    c. 170 > 125   d. 400 < 404

16. a. 83
    b. 682
    c. 748

17. a. 24, 61
    b. 722, 970

18. The total cost was $304. 152 + 152 = 304

19. 450 − 126 = 324 DVDs

20. 218 + 218 = 436

21. a.

300 yards

150 yards   150 yards

300 yards

    b. 300 + 300 + 150 + 150 = 900 yards

22.

a. 3 : 25 25 past 3	b. 8 : 50 10 till 9	c. 4 : 35 25 till 5

23.

from	3:00	2:00	1 AM	11 AM	8 PM
to	3:05	2:30	8 AM	1 PM	midnight
amount of time	5 minutes	30 minutes	7 hours	2 hours	4 hours

24. a. $1.30   b. $5.41

25. The change is $0.65.

26. His change was $0.15.

27. Shape A: <u>a square</u>   Shape B: <u>a pentagon</u>

28. a.

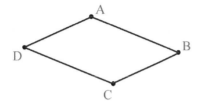

    b. a quadrilateral (also acceptable: a parallelogram)
    c. Side AB: about <u>about 2 inches</u>   Side BC: <u>about 1 1/2 inches</u>
       Side CD: about <u>about 2 inches</u>   Side DA: about <u>about 1 1/2 inches</u>

29. about 9 cm

30. It is enough for the student to mention one suitable measuring unit. All the units mentioned below are right answers.

Distance	Unit(s)
how long my pencil is	cm, in
the distance from London to New York	km, mi
the height of a wall	m, ft
the distance it is to the neighbor's house	m, ft

31.

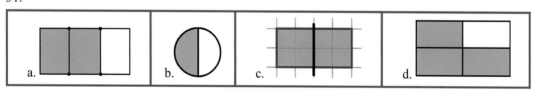

32.

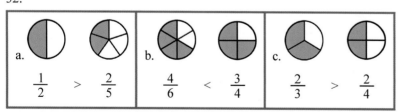

33. a. $2 \times 3 = 6$ b. $6 \times 2 = 12$

34. a. $3 \times 5 = 15$ b. $5 \times 4 = 20$

35. a. 10   b. 9   c. 30

Made in the USA
Las Vegas, NV
12 May 2021